Diplomatic Families and Children's Mobile Lives

This book is the first of its kind: a historical inquiry into the family life of British diplomats between 1945 and 1990. It examines the ways in which the British Diplomatic Service reacted to and was influenced by the radical social changes that took place in Britain during the latter half of the twentieth century. It asks to what extent diplomats, who strove to protect their enclosed and elite circles, were suitable to represent this changing nation.

Drawing on previously unseen primary sources and interview testimony, this book explores themes of societal change, end of empire, second wave feminism, new approaches to childcare, and developments in the civil service. It explores questions of belonging and identity, as well as enduring perceptions of this organisation that is (often mistakenly) understood to be quintessentially 'British'.

Offering new and fresh insights, this book will be of interest to students and scholars in history, historical geography, political studies, sociology, feminist studies and cultural studies.

Sara Hiorns is a historian of modern Britain with a special interest in the social and cultural history of the British Foreign Office.

Routledge Spaces of Childhood and Youth Series
Edited by Peter Kraftl and John Horton

The Routledge Spaces of Childhood and Youth Series provides a forum for original, interdisciplinary and cutting edge research to explore the lives of children and young people across the social sciences and humanities. Reflecting contemporary interest in spatial processes and metaphors across several disciplines, titles within the series explore a range of ways in which concepts such as space, place, spatiality, geographical scale, movement/mobilities, networks and flows may be deployed in childhood and youth scholarship. This series provides a forum for new theoretical, empirical and methodological perspectives and ground-breaking research that reflects the wealth of research currently being undertaken. Proposals that are cross-disciplinary, comparative and/or use mixed or creative methods are particularly welcomed, as are proposals that offer critical perspectives on the role of spatial theory in understanding children and young people's lives. The series is aimed at upper-level undergraduates, research students and academics, appealing to geographers as well as the broader social sciences, arts and humanities.

After Childhood
Re-thinking Environment, Materiality and Media in Children's Lives
Peter Kraftl

Why Garden in Schools?
Lexi Earl and Pat Thomson

Latin American Transnational Children and Youth
Experiences of Nature and Place, Culture and Care Across the Americas
Edited by Victoria Derr and Yolanda Corona-Caraveo

Mapping the Moral Geographies of Education
Character, Citizenship and Values
Sarah Mills

For more information about this series, please visit: https://www.routledge.com/ Routledge-Spaces-of-Childhood-and-Youth-Series/book-series/RSCYS

Diplomatic Families and Children's Mobile Lives

Experiences of British Diplomatic Service Children from 1945 to 1990

Sara Hiorns

Routledge
Taylor & Francis Group

LONDON AND NEW YORK

First published 2022
by Routledge
2 Park Square, Milton Park, Abingdon, Oxon OX14 4RN

and by Routledge
605 Third Avenue, New York, NY 10158

Routledge is an imprint of the Taylor & Francis Group, an informa business

British Library Cataloguing-in-Publication Data
A catalogue record for this book is available from the British Library

Library of Congress Cataloging-in-Publication Data
A catalog record has been requested for this book

ISBN: 978-0-367-22164-5 (hbk)
ISBN: 978-1-032-12025-6 (pbk)
ISBN: 978-0-429-27356-8 (ebk)

DOI: 10.4324/9780429273568

Typeset in Bembo
by Taylor & Francis Books

Contents

Abbreviations

DSA	Diplomatic Service Association
DSWA	Diplomatic Service Wives Association
FCO	Foreign and Commonwealth Office
FSA	Foreign Service Association
FSWA	Foreign Service Wives Association
PUS	Permanent Under Secretary (the head of the Diplomatic Service)
TCK	Third Culture Kids
TNA	The National Archives, Kew
UAF	Unaccompanied Air Freight

Introduction

In October 2013, a magazine article about British punk band The Clash had this to say about the band's lead singer, Joe Strummer:

> Punk fundamentalists were ... outraged when they discovered that Joe had gone to some minor private school ... suddenly, he was inauthentic, a poseur playing at being a punk, not the real thing.[1]

These elements in Strummer's family background, at odds with the iconoclastic, aggressive nature of the punk movement, have been widely reported. During his lifetime Strummer was at great pains to neutralise the assumption that he had been raised as a member of the establishment. In interviews he gave poignant accounts of an unhappy spell at boarding school, and spoke about his father's job – which paid the boarding school fees – with derision.[2] Joe Strummer – or John Grahame Mellor as he was born in 1952 – attended a private boarding school because his father Ronald worked for the British Diplomatic Service, the branch of the civil service that represents British interests overseas. The two-dimensional and dismissive criticism that he faced was not an unusual reaction to the assumed circumstances of diplomatic children from their contemporaries.

More apposite, however, to the genuine experience of diplomatic children is a comment about Strummer 'trying on aliases for size' because he struggled to find a personal identity.[3] Another article stated that 'beyond ... a worldly background ... he didn't have a lot to give'.[4] The influence of world music that became a familiar part of The Clash's sound has been ascribed to a receptiveness to other cultures that was the result of Strummer's peripatetic early years.[5] A former partner commented on the contradictions in his personality: 'Joe was charming. He was brought up as an embassy child – he knew how to fix you a drink. But he was not all charm: he was a chameleon – he was exactly what you wanted him to be.'[6] Strummer's relationship with his parents was ambivalent. They got on, despite talk of neglect and heartlessness when he spoke about boarding school. Yet one subject he remained reticent about was the suicide of his brother David – another Diplomatic Service child – in 1970. Joe Strummer achieved worldwide fame as a musician, and the ways in which his formative experiences shaped his personality and his subsequent engagement with British society at a time of rapid change have been foregrounded in interviews and biographies. However, these basic themes of family separation, the demands of a transient lifestyle, the pursuit of a stable identity and a sense of belonging are as evident in accounts of his life as they are in the accounts of all the diplomats' children who grew up during the second half of the twentieth century and whose lives – and those of their families – are explored and documented in this book.

This book is the first historical reconstruction of the lives of British Diplomatic Service families between 1945 and 1990 that concentrates on the mobile lives of their children and is based on a wide range of sources and approaches. Although primarily historical, it has been necessary to read across disciplines in order to make sense of the often singular lives of diplomatic families which are highly influenced by the institution that employs one or more parents. Its intention is to elucidate the distinctive diplomatic communities – which were often at odds with the contemporary society they were meant to represent – in which these children grew up and to attempt an analysis of their social and cultural habits. The narrative examines the lives of British Diplomatic Service families at time of rapid and significant social change in the UK and establishes how the story of these families between 1945 and 1990 intersects with many narratives within the history of children and family life, for example, those of colonial service or 'Raj orphans', those of public school children or even those of refugees, but it will demonstrate how they were, at the same time, distinct. This is perhaps best seen in the way that the experiences of Diplomatic Service children began, early on, to diverge considerably from those of their contemporaries at 'home' in Britain.

The demands of Diplomatic Service life can be understood through the lens of 'total institution' – one which controls, influences and imposes an irresistible structure on its employees and their families. Within this, various reoccurring factors regulate many aspects of its employees' lives, including an emphasis on elite social class and highly gendered modes of behaviour. As will be seen, however, the most conspicuous element of this book is its documentation of the Diplomatic Service's determined resistance to change, despite attempts at every level to force this. Another compelling aspect of diplomatic life is the ambiguity often inherent in its structures and attitudes and this book is no exception. The Foreign Office's resistance to change is explored, but also touched on is the apparent contradiction that while Foreign Office children seemed to be, in common with the institution itself, 'behind the times', they were in many ways at the forefront of a lifestyle which has stimulated related writing from handbooks on 'Third Culture Kids' to the work of postmodern anthropologists and scholars of transnational migration. A study of diplomatic children provides a new focus for students of diplomatic history and the social history of the structure and function of the Diplomatic Service. Despite the fact that their early lives were dictated by the Conditions of Service set down by an organisation perhaps over-fond of regulations, very little about children can be found in existing literature on the Foreign Office. Looking, then, at the Foreign and Commonwealth Office (FCO) through the lens of children and families allows us to rethink familiar narratives in its institutional history.

Conceptual background

Concepts of what it means to be a child and the meaning of childhood have differed markedly over time. Yet despite the presence in every society of a group of people in the early stages of the life course, the study of children's history was late to develop. 'As late as the 1950s,' Colin Heywood has noted, 'their territory could be described as "an almost virgin field"'.[7] *Centuries of Childhood* published by Philippe Ariès in 1962 is acknowledged as the first study in this field. While Ariès' claim that 'in medieval society the idea of childhood did not exist' drew criticism from historians who questioned the (in their view) insubstantial and idiosyncratic quality of his evidence, others embraced his thesis, showing specific interest in identifying the moments in history (the 'turning points' as they have become known) when attitudes towards children changed.[8] The lively exchanges provoked by Ariès' work perhaps explain the exponential growth of this area, although the 'reformation' of the family from the mid-century onwards which gave children in western societies a central place and role in family life could also be responsible. Since the publication of *Centuries of Childhood*, the history of childhood has diversified to the extent that Peter Stearns has recently been able to contextualise it in global terms.[9] Since historians began to conceptualise childhood as a social 'construct', new avenues of research have opened up exploring the intersections of childhood with gender, race, class and other categories of difference.[10]

It is possible to analyse the experiences of Diplomatic Service children, who were well travelled, culturally aware and whose parents – or fathers at least – were engaged in international careers, in terms of a global history of childhood. However, the lives of Diplomatic Service children during the timeframe under consideration (1945–1990) speak to many different contexts, including, for instance, the considerable body of work on colonial children.[11] But, the Diplomatic Service was distinct from the colonial service in important ways. The diplomatic mission, or embassy, existed to represent the British government and monarch overseas while colonial administrators directly oversaw the colony's government.[12] It was rare that colonial administrators served in more than one country, unlike diplomats who changed jobs every two to three years. Social practices, however, such as the tendency to educate children at boarding schools in Britain, were shared amongst diplomats, colonial administrators and the wider British expatriate community. Family history and sociological studies of the family are also useful in order to gain an understanding of the changing lives of children and the positionality of family members, but these very rarely (if at all) encompass the additional strains placed on a family required to move regularly and to adapt to life in different countries and cultures.[13] Gittins' description of childhood history as a 'mass of tangled strands' is particularly suitable here.[14] For, in addition to finding a context within global history, the experiences of Diplomatic Service children might as easily relate to 'transnational' family studies or to the study of migration and diaspora.

The application of 'transnationalism', a social phenomenon describing movement of people, goods and ideas across national borders that gained currency among anthropologists in the 1990s, has provided a useful setting for the more complex and abstract concepts explored in a study of British diplomats' children, especially those surrounding identities and ideas of home.[15] The work of Blunt has been key to the development of thinking about home in a transnational context.[16] Chambers' *Migrancy Culture Identity* written in 1994, which explores the implications for the construction of a self-image amidst temporal and spatial flux, offers a plausible framework for analysing the conflicting expressions of identity explored by Diplomatic Service children. Crucially, it should be noted that although these conceptual tools were developed by scholars seeking to capture the complex processes of globalisation in the 1990s, the phenomena they describe and analyse were present earlier in the lives of transitory diplomatic children.[17]

An inherent difficulty with placing the history of Diplomatic Service children within the transnational context of global movement is the marked difference in status between the migrants featured in the global studies literature and those studied in this book. With the exception of specific works on British colonial children, like those of Buettner and Pomfret, for the most part the groups discussed in the literature are non-western economic migrants or groups who have been compelled into becoming migrants: refugees, for example, or those who were literally forced into movement by the transatlantic slave trade. The identification and use, then, of the term 'transnational professionals' whose family lives provide the topics for the papers collected by Anne Coles and Anne-Meike Fechter in *Gender and Family among Transnational Professionals* is far more appropriate.[18] Coles' chapter takes the findings of a survey of FCO family life conducted in 2004 as its subject. This significantly postdates the time frame covered in this book but illustrates the changes and continuities in diplomatic culture. Also pertinent to an exploration of Diplomatic Service children in the same collection is Moore's exploration of the pupils and parents of the German school in London, which considers not only the way in which children at school overseas relate to their host environment and to one another but the ways in which parents work to construct and thus take pride in their children's sophisticated international personas.[19]

This satisfaction exhibited by transnational parents in the cosmopolitan sophistication of their children is a theme that emerges in other works by social anthropologists which have great resonance for the study of British Diplomatic Service children. Roger Goodman's study of *Japan's International Youth* deals with the experiences of Japanese children (known in Japanese as *kikokushijo*) removed from their native culture to travel with their parents whose employment (usually by the well-known Japanese multinational companies who enjoyed significant growth during the late twentieth century) took the whole family overseas.[20] The *kikokushijo* can be linked to another area of social anthropology that has gained immense popularity with international families, especially those in the US: that of the 'third culture' which gave rise to the phenomenon of 'Third Culture Kids' and which will be discussed at length in later chapters.

Contexts

Examining the British Diplomatic Service family through a transnational lens and making use of an interdisciplinary approach offers a useful set of conceptual tools with which to make broader sense of the experiences of children in the British Diplomatic Service. The focus of this book, however, is historical and it is necessary to contextualise the position of the British diplomatic family within the historical scholarship concerning the period immediately succeeding World War Two. Our main line of inquiry lies at the intersection of two key literatures, one exploring change and continuity in family life in post-war Britain, and the other analysing the shifting nature of Britain's global role and the place of the Foreign Office and Diplomatic Service within it. The literature on childhood and the family in the post-war decades explores a number of tensions: for example, the way in which the family inhabited the public and private spheres and the roles fulfilled by individual family members and the relationships between them. Again Gittins' 'tangled mass' analogy is germane here.[21] To gain an understanding of the challenges facing any family in this period we first need to understand the changes taking place in post-war society. Study of the *diplomatic* family, however, must also be conducted in terms of its deep-seated connection to an exclusive group within British society, the Diplomatic Service. In contrast to British families in general, the history of the Diplomatic Service family has not been widely researched. Diplomatic wives have received some attention, but children have not been treated in the same way.[22]

As will become clear, diplomatic families cannot be easily located in existing narratives of affluence and home-centredness which dominate accounts of family life in post-war Britain.[23] The finer points of daily life in the UK were something of which members of diplomatic families were often largely ignorant. Some diplomatic wives reported that they had no idea where to make a claim for child benefit on their return to the UK and children were ostracised at school because they were unfamiliar with the most recent television programmes. However, there can be no doubt that the wider social shifts that changed the lives of Britons from 1945 onwards also influenced those of the Diplomatic Service family, although in complex ways. Changes in the nature of Britain's social structure after the upheaval of the Second World War, combined with the new system of education introduced by the 1944 Education Act, gave young men from the working and lower middle classes the opportunity and confidence to apply and gain acceptance into the Diplomatic Service. 'Chairmen of the largest industrial companies, top civil servants, and churchmen between 1880 and 1970 were increasingly drawn from lower part of the middle class ...' writes Harrison.[24] Whether these 'meritocratic' recruits remained true to the characteristics of their own 'class', however that category was understood at the time, or were pleased to move upwards into a world of greater privilege – especially within the traditionally elite Foreign Office – is a question to which this book returns and attempts to answer.

By 1975, significant trends that were re-shaping the lives of women in Britain also affected the Diplomatic Service family. Denise Riley has shown how differences between conceptions of *women* as workers and as *mothers* were inscribed in the post-war welfare state, contending that post-war anxieties about children's war experiences (of evacuation and blitzkrieg) combined with a rise in popular psychology, especially the theories of Donald Winnicott and John Bowlby, led to women with young children being discouraged to seek employment.[25] Nonetheless, increasing numbers of married women sought paid work and by the mid-1970s 'more British women were returning to paid work after having children than in any other EEC country'.[26] These conflicting dynamics are evident from a survey of the magazine produced by the Diplomatic Services Wives' Association which underlined the fact that diplomatic wives began to question their life of voluntary good works and official entertaining at this time and to agitate for acknowledgement (and remuneration) from the Foreign Office and for opportunities elsewhere.[27]

But how did developments affecting families in British society during the twentieth century – 'the century of the child' – affect British children, and diplomats' children more specifically?[28] The tension between child-centred methods of child-rearing (which had rapidly entered the public consciousness as the 'correct' way) and the long-established patterns of separation which British colonials claimed were necessary to raise their children will be one of the key themes of this book. Cunningham describes the high hopes invested in children in the late 1940s during the founding of the Welfare State, the relaxation in parental discipline and, linked to this, the way in which the child became central to the family, stating that parents' hopes 'became inseparable from the happiness of and success of their children'.[29] In a major study published in 2013 Mathew Thomson explored two common themes in the history of childhood: those of a loss of childhood freedom and the rise in children's rights. Thomson sought to explain the change in attitudes that led to children's lives becoming more constrained; he considers the 'landscapes' of childhood, how they became centred on home and how this was made possible within the structure of the 'post-war settlement'.[30]

On the global movement of children during the twentieth century, studies exist in many areas. For example, Boucher's *Empire's Children* explores the sponsored migration schemes that took British children to remoter parts of the empire and later to Commonwealth countries from the late nineteenth century until the middle of the twentieth.[31] The period directly preceding the Second World War and the war years themselves were also characterised by children on the move and British children – especially city children – were introduced to children from outside the UK for the first time. Among these, for example, were the four thousand Basque children who were allowed entry to the UK seeking sanctuary from the civil war in Spain in 1937. They acted as forerunners to the far better known *kindertransports* which brought Jewish and other children from Germany, Austria and Czechoslovakia whose cultural or religious beliefs put them in danger of Nazi persecution.[32] During the war years British children themselves experienced profound changes of circumstance when government evacuation schemes removed city children into the countryside to escape the effects of aerial bombardment. Some children went further afield – to the Commonwealth – as a result of the Children's Overseas Reception Board (CORB) scheme.[33] B.S. Johnson, who collected the memories of a group of former evacuees in 1968, revealed the feelings of bitterness and sorrow maintained by many. Many of the evacuees' recollections find an echo in the interviews conducted with former diplomatic children and quoted in this book. The children were not so much angered by the family separation they endured as by the expectation that life would return to normal when families were reunited. They expressed the collective opinion that an irreparable gulf of understanding had formed between the adults and children. The separation experienced by these families bears similarities to the experiences of Diplomatic Service children, for instance in the ways evacuees struggled to adjust to 'normal' life when the war was over.[34]

Diplomatic children and family structures are, as has been already suggested, only sparsely covered in an area of scholarship which centres on the political and administrative development of the Diplomatic Service rather than its cultural or social aspects. This book is the first of its kind but there are some forerunners, which have been a very useful foundation on which to build. Jennifer Mori devoted a chapter of her study of British diplomatic culture in the earlier period of the late eighteenth and nineteenth centuries to 'Family, Sex and Marriage'.[35] An important chapter that introduces many of the issues faced by Diplomatic Service families, as well as a wider consideration of Foreign Office culture, can be found in *Daughters of Britannia* written in 1999 by Katie Hickman on the history of diplomatic wives.[36] Hilary Callan's *Premiss of Dedication Notes on the Something of Diplomatic Wives* was a very necessary addition, as was the volume that she co-edited almost a decade later with Shirley Ardener, *The Incorporated Wife.* [37] Steiner and Cromwell made two significant contributions to the slender body of objective work on Foreign Office society and culture, on which I have drawn extensively, despite these both being relatively short.[38] Also most pertinent have been Coles and Fechter's *Gender and Family among Transnational Professionals*; Helen McCarthy's 2014 monograph, *Women of the World*; and the work of Glenda Sluga and Carolyn James in this area which – while not dealing directly with the experiences of children – offers useful insights into how a history of the 'unofficial', 'informal' and intimate aspects of diplomatic life can be integrated into the mainstream narratives of the FCO and Diplomatic Service.[39]

Both Elizabeth Buettner and Vyvyen Brendon have written extensively on the children of Anglo Indians (i.e. those Britons who administered, traded and spread the gospels in Imperial India). Their work reveals many parallels between diplomatic and colonial children. This is perhaps why the two groups remain synonymous in the British popular imagination. While less substantial and more popular than Buettner, Brendon nonetheless brings the discussion up to date in a number of areas. She touches for example, on the often poignant and sometimes tragic, continuities – past and present – of informal international fostering and covers the growing awareness of 'international children' of their own rare lifestyle.[40]

For diplomats' children boarding school was often the most significant aspect of family separation. Brendon, writing about prep schools in 2009, points out that they 'are an area of childhood hardly mentioned in recent histories ...'[41] Boarding schools appear in this book both as a part of children's experience and a societal phenomenon within British life. Although boarding schools existed as part of the state education system, they are often viewed as interchangeable with public schools; as Millham and Lambert put it: 'To most people "boarding education" is simply equated with boarding school education.'[42] They were frequently linked, during the period under consideration, to ideas of nation, empire and class, very much in the same way the Diplomatic Service was linked to imperial values and traditions. Research on boarding and public schools between the wars emphasises their strong sense of ritual and the demands of unquestioning devotion from pupils, who often retained lifelong ties with the school and established a generational identity, thus, again, comparable to the Diplomatic Service.[43] Writing in 1967, historian of education Bamford observed that Old Boys' societies produced 'an immense literature, almost wholly devoid of objectivity and analyses'.[44] Callan uses Goffman's concept of 'total institution' to interpret this absolute loyalty.[45]

Research background

The decision to focus this study on the years 1945 to 1990 was made for a number of reasons. First was the sense that there was a great urgency to record recollections from diplomatic children who were still living and who could recall their experiences from the early post-war period.[46] Second was the additional amount of training and administration required to interview children and young people under the age of sixteen. The interviewer would have had to have undergone a Criminal Records Bureau check and become familiar with interview techniques developed for children, which while all perfectly possible would have been time-consuming. Also, parents would have been a required presence at the interviews, a practical difficulty and additional time concern when dealing with busy civil servants working across a global network. Given that this study was conceptualised as historical enquiry rather than an exploration of contemporary children's lives, the decision was taken to focus on adults whose diplomatic childhoods formed part of their pasts. The risk would have been that this extra level of preparation might not have yielded results of sufficient quality or quantity.[47] Once a substantial amount of research had been covered it was clear that 1990 was a very suitable end date. Changes in the circumstances of diplomatic children in the last twenty-six years are touched on in the conclusion, but it is hoped that this period will become the topic of further study.

The historian must be willing to maintain a critical distance from their subject. However, the circumstances behind this book are slightly unusual because I was employed from 2004 until 2016 by the FCO, the institution which my research explores. It is important, therefore, to acknowledge what social anthropologists would call my 'insider-outsider' status.[48] This is not to say that, through personal involvement with the institution, I hold superior knowledge or understanding, but rather that my perspective as a researcher contains a tension between what Kirsten Haastrup has described as the insider's 'implicit' knowledge of a situation and the 'explicit' knowledge of the outsider.[49] I have striven, at all times, to question my assumptions and guard against allowing my own subjective experiences of working in the Office to shape the analysis of the historical evidence, although I am aware that total objectivity is impossible to achieve, particularly on the emotive topic of family. An ability to, in Leonie Gordon's words, 'figuratively step outside' the area of study is essential if patterns of behaviour are not going to be taken for granted or simply go unnoticed.[50] However, as Ann Coles has observed, the FCO is a secretive organisation and 'insider' status can be very helpful in putting FCO participants at their ease, for instance during interview.[51] Coles herself experienced life in the Foreign Office because she was married to a diplomat, this she has in common with Hilary Callan, while Katie Hickman is the child of a diplomat. In practical terms, a knowledge of the FCO's somewhat arcane staff structure and use of language saved a great deal of time in the project's initial stages. It should be added that – for variation – the terms 'Diplomatic Service' and 'Foreign Office' are used interchangeably in the text.

A question central to a thesis which aims to reconstruct the lives of diplomats' children in the past is that of how a 'child' is defined. Age seems to be the most obvious marker: as Cunningham writes, certain 'biological imperatives' are difficult to ignore.[52] However disparate, many societies maintain 'coming of age rituals' as young people enter adulthood. P.J. Rich describes the convention of young English boys graduating from prep to public school as 'a *rite de passage* as marked as a ritual of aboriginals or a calamity of the wild'.[53] The experience of secondary education is one distinguishing feature of western childhood. Ariès proposed that childhood maintained as a distinct state from other age groups in society was a defining condition of modernity, and many of the qualities we would name as characterising childhood and youth in early twenty-first century Britain – those of adulthood long deferred and extended periods of education, as well as the expectation of happiness and assumption of innocence and need for protection – are constructs of the culture within which we live.[54] An interesting view on the sometimes impenetrable nature of childhood is expressed by Peter N. Stearns, who reminds his reader 'we have all been children so we know the topics involved at least to some extent …' and we should not forget that the definition of a child is relative: even as adults with children of our own, we remain the children of our parents.[55] It is important to keep in mind that the 'diplomatic children' who actively engaged in this project, either through taking part in interviews or responding to questionnaires, were adults recalling their childhoods. Perhaps, then, 'diplomatic children' are best described for the purpose of this book as individuals who were at varying stages of the lifecycle but whose parents were employed by the British Diplomatic Service and who spent time as children within the structure and conventions of the British Diplomatic Service between 1945 and 1990.[56]

Having to rely on imperfect memories, which can be obscured by hindsight or sentiment, illustrates the difficulties regularly experienced by historians of childhood in their search for primary sources. Addressing the analytical challenges of writing children's history, Stearns admits that children simply do not produce the same number of sources as adults.[57] Hendrick emphasises that evidence relating to British children is heavily dependent on social class. For example, during the modern period, social investigators tended to concentrate on working class children to gather evidence for projects concerned with poverty, juvenile labour, criminality and other social problems. Very little documentary evidence came directly from the working class participants themselves, however.[58] In the nineteenth and early twentieth century, middle and upper class children might have produced their own documents, such as diaries and manuscript magazines, but these give only a very narrow impression of the lives of a small group of children.[59] One such manuscript source forms part of this book: *The Double J,* a magazine that was authored by John and Jennifer, the children of well-known diplomat, Pierson Dixon, between 1941 and 1948. *The Double J* gives a very specific view of the life of a tight-knit, upper middle class intellectual family and the dominance of the Foreign Service – as it was at that time – as an integral part of the Dixons' family culture is evident from its presentation and articles.[60]

The reader cannot ever be sure of the extent to which written evidence that has apparently been produced by children has been subject to adult intervention. Some examples of children's writing are undoubtedly initiated by adults in order to 'showcase' the talents of their children. During the 1980s an acknowledgement of children as part of the FCO community (or perhaps due to a scarcity of articles written by adults) led to requests for FCO children to submit written pieces for the Diplomatic Service Wives Association (DSWA) newsletter. The first spate of children's articles, encouraged to coincide with the UN's International Year of the Child in 1979/1980, were published under the heading 'Views of parents and children' and offered a broad range of opinion on the children's lives, some by parents.[61] A request for further writing from children was published in the Autumn 1982 edition, urging FCO youngsters to 'write about any aspect of FCO life that interests you'.[62] The magazine continued to publish occasional batches of children's writing and, as the 1980s progressed, could also be said to have become a space where the different generations of FCO families expressed their opinions about one another.

Heywood has cautioned the historian of childhood, to whom so little direct source material is available, to remember that all discourses must be subject to interpretation and not taken at face value.[63] While Stearns accepts caution is necessary he writes of the pleasures of 'analytical agility' open to the childhood historian working across disciplines and variables.[64] Historians of childhood recognise a need to interpret sources and to consider the preoccupations of those who produced or commissioned them. We need, therefore, to acknowledge that the many articles about diplomatic children that appear in the DSWA magazine – predominately with regard to education at all levels and travel, although health and disability are also touched on – are valuable sources but that they were no doubt 'mediated' through a selective editorial process.

That the bulk of concerns for children were expressed by diplomats' wives during this period points to highly gendered and traditional roles upheld within the structures of diplomatic marriage and family.[65] This study has been difficult to situate within established historical narratives of the Diplomatic Service. This is because, despite the fact that the FCO has a team of historians working in White-hall, their primary function is to maintain fields of scholarship which centre on the political and administrative work of the Foreign Office. Thus within the FCO itself very little scope exists to promote investigation into the social and cultural aspects of diplomatic life. The diplomatic family, although closely involved in diplomatic work during the period covered in this book, has never been considered appro-priate as a subject for research. This one-sided culture has led to the rich sources stored by the FCO historians being under-used and to uneven availability of material at the National Archives. It seems likely that many papers were not deposited there because they were not considered sufficiently important.

It is clear from the papers kept by the FCO historians in Whitehall relating to Beryl Smedley's 1990 book *Partners in Diplomacy* that care of diplomatic children was understood as 'women's work'.[66] The Smedley Papers include the original questionnaires completed by wives which formed the basis of Smedley's book, alongside a collection of newspaper cuttings on diplomats and diplomatic life as well as issues confronting women during the 1980s. Other memoirs by wives concentrate on the domestic travails of the diplomatic family and usually contain a chapter or section devoted to diplomatic children. Some wives' memoirs are very obviously aimed at the popular market and self-consciously set out to provide a light-hearted and entertaining account of diplomatic life, although Coles observes that writing about their unusual lifestyle, and writing generally, is a convenient and portable career for a woman married to a husband with an international career.[67] A far greater number of memoirs were written by male diplomats, but they con-tain far less information about their children and it is necessary to read more of them to piece together a picture of FCO family life from a male – or father's – point of view. The existence of what Walsh refers to as the 'persistent gendering of expatriate lives' is detectable in the apparent distance between male diplomats and their children and between the gendered roles of diplomatic parents.[68] If the 'overt' public sphere of diplomatic and political work is masculine, then children – whose place was in the 'covert' feminine sphere – existed only on its margins.

Official FCO documentation relating to families is uneven, which could be because its preservation was not considered as a high priority. However, several files kept at the National Archives enable the researcher to reconstruct official attitudes towards families and the development of institutional policies which affected them. Most useful for the early section of the period are Conditions of Service documents which summarise the financial allowances that were extended to diplomatic families and most regularly detail children's travel costs and boarding school fees, the two areas most diplomatic families commonly cited as problems.[69] FCO files also review nationality issues concerning Foreign Office children born overseas, healthcare provision for children at post and a large amount of correspondence exists (dated 1971–1972) concerning a third 'concessionary' (or subsidised) journey for children to visit parents overseas.[70] Another rich official source is the 1964 Report on Representational Services Overseas, known as the 'Plowden Report' after its chair, Lord Plowden.[71] The Plowden Committee papers contain direct evidence from diplomatic staff and their wives on the subject of family and have been crucial to this study.[72]

The recognition of oral history as a legitimate and valuable research methodology has grown concurrent with the time frame covered in this project, described by Perks as enjoying a 'post-Second World War renaissance'.[73] Oral history has been useful, as Thomson suggests, 'by bringing recognition to substantial groups of people who had been ignored'.[74] Oral history greatly shaped the development of women's history and that of other groups outside the mainstream and has had a 'transforming impact' on the history of the family.[75] From the outset, it was envisaged that oral history would add an essential dimension to the sources already named and the initial aim was to interview between thirty and forty former diplomatic children, which seemed an appropriate number given the constraints of working as a single researcher. Adverts were placed in a weekly online bulletin, internal to the FCO, and on websites maintained by the Diplomatic Service Families Association and the FCO Association for retired members of the Diplomatic Service. An initial surge of interest gave a misleading impression of willingness and availability, however, and ultimately only eighteen interviews took place (see Appendix 1 for brief biographical details of interviewees and correspondents). These were semi-structured life interviews that took place at a location of the participant's choice – usually in their own home. Ahead of the interviews participants were sent a pro-forma information sheet which included a suggested schedule of the kinds of questions that would shape the conversation (see Appendix 2).

It is useful to note that there was a suspicion among many of the interviewees that the specific intention of this project was to reveal memories of traumatic or difficult childhoods and present these as a direct result of a childhood spent in the Diplomatic Service. Contributors sometimes anticipated this by making a joke about 'skeletons in the closet' as the interview began: one member of a large family immediately took the subject of the project to be 'damaged diplobrats' and communicated this to her siblings who wrote to me separately quoting the term more than once, although it had never been mentioned anywhere else. When it came to writing up the book I offered all correspondents the opportunity for their contribution to be anonymised. In the light of the foregoing sentences, a surprising number were happy to be named and their real names appear in the text. Those who preferred to remain anonymous have been given a pseudonym which is clearly indicated with an asterisk. I decided to do this because I felt that names, rather than initials or symbols, keep the text flowing.

The sense of unease and suspiciousness indicated by contributors is difficult to unravel, and may have roots in loyalty to the Foreign Office as an institution, or loyalty to parents who belonged to that institution. In contrast to the work of the British Diplomatic Oral History Programme, which provides a research resource focussed on the formal careers of British diplomats and their contribution to foreign policy, the complex and demanding nature of talking to an investigator about emotional, personal and family circumstances should not be underestimated.[76] This lack of certainty and feeling of exposure became evident when the idea of a witness seminar involving interviewees and academics at the Foreign Office in London was proposed. The suggestion received a reaction which amounted to a sense of grievance or breach of trust among interviewees and had to be abandoned. The formal academic process of the proposed seminar, held in the imposing surroundings of Whitehall, sat uneasily with the highly individual expressions of childhood emotions, a tension foregrounded by Linda Shopes in her writing about community oral history.[77] The level of sentiment expressed *within* the interviews, however, is a different matter and can be interpreted, like the work of Stoler and Strassler on Dutch colonial children, as a narrative on the self: a version constructed from 'scripted, storied narratives'.[78] Placed alongside the other sources outlined above, however, a reconstruction of the diplomatic child's experiences are possible using oral testimonies. These are available, as Hendrick has written, 'via whispers and muted articulations, albeit … in the form of adult recollections'.[79]

Chapters

This book began life as a doctoral thesis, and in that incarnation its chapters were themed. It has been reworked along chronological lines, to encompass the main milestones of diplomatic family life. Owing to the 'patchy' and unbalanced nature of evidence available, these chronological chapters cover long and seemingly arbitrary time spans. They were, however, carefully chosen – for example to illustrate 'before and after' watershed moments, or to group relevant events together – and the reasons for this will be given within each chapter outline below.

Chapter One: 1945–1958

This chapter provides an introduction to the Diplomatic Service family in the years after the Second World War. The time frame was specifically chosen to begin with the immediate post-war period and end just as a small group of diplomatic wives were considering how to establish improved pastoral and welfare systems. It begins with a section illustrating the social demands made of diplomats and the subtle ways in which they enforced their closely guarded hierarchy. It then moves to an analysis of the circumstances in which the British family found itself in post-war Britain. The final section is concerned with an introduction to the most compelling aspect of Diplomatic Service family life during the whole period under consideration, that of long periods of family separation.

Chapter Two: 1958–1971

Chapter Two covers the period from the formation of the Foreign Service Wives Association (FSWA), a landmark in the social history of the Foreign Office, to the granting of a Third Concessional Journey for children's travel to and from overseas post. It begins with a section on the Plowden Report, to whose Committee both members of the Foreign Service Association and the newly established Foreign Service Wives' Association gave evidence and whose reforms made such a difference to the lives of diplomats' families. The second section considers the Foreign Office in the light of the new social mores that characterised 'the swinging sixties' and asks why Diplomatic Service children appeared reluctant to rebel. The final section looks in detail at the campaign for finance for a third journey so that children could visit their parents at post for every school holiday.

Chapter Three: 1972–1985

The third chapter begins with the period which saw the Foreign Office marriage bar, which forced women diplomats to resign on marriage, lifted. This – along with social changes in the UK linked to second wave feminism – led to a shake-up of attitudes in the Diplomatic Service, especially among wives, whose discontent and its expression is outlined in the first section. This period also saw a rise in the level of threat towards Foreign Office families overseas and the second section looks at two specific cases of serious threat experienced by families. The final section includes, for the first time, the voices of Diplomatic Service children, as they appeared in the Diplomatic Service Wives Association (DSWA) magazine in 1979 and 1980 in conjunction with the UN International Year of the Child.

Chapter Four: 1985–1990

Chapter Four is an examination of the ways in which Diplomatic Service family life had remained static and the ways in which it had changed. The first section examines an institution apparently stuck in the past and held fast by its own traditions – family separation, still the normative experience for families in this period is revisited. The second section, however, looks at ways in which the Foreign Office, and, by association, the experiences of its families, was beginning to change. It shows how attitudes were beginning to be influenced by rapid changes outside the Diplomatic Service. The final section looks towards the future and examines the ways in which the children of transnational professionals and their parents began to seek new labels and identities in the post-colonial world.

Notes

1 Alan Jones, 'The Clash Live On!', *UNCUT Magazine,* October 2013.
2 Chris Salewicz, *Redemption Song: The Definitive Biography of Joe Strummer* (London: Harper Collins, 2007), 47–48.
3 Clinton Heylin, *Babylon Burning: From Punk to Grunge* (London: Penguin, 2008), 61.
4 J.H. Tompkins, 'Joe Strummer's Complex Legacy: 10 Years After the Clash Leader's Death', *Spin Magazine*, December 2012.
5 Barry J. Faulk and Brady Harrison, eds, *Punk Rock War Lord: The Life and Work of Joe Strummer* (London: Routledge, 2016), 5.
6 Salewicz, *Redemption Song*, 245.
7 Colin Heywood, *A History of Childhood: Children and Childhood in the West from Medieval to Modern Times* (Cambridge: Polity Press, 2001), 5.
8 Phillippe Aries (trans Robert Baldick), *Centuries of Childhood A Social History of Family Life* (London: Jonathan Cape, 1962), 128; for a detailed discussion of Aries, his critics and supporters see Heywood, *History of Childhood*, 9–13.
9 See Peter N. Stearns, *Childhood in World History: Themes in World History* (London: Routledge, 2011).
10 See Paula S. Fass, 'Is there a story in the history of childhood?' in *The Routledge History of Childhood in the Western World*, ed. Paula S. Fass (London: Routledge, 2013), 1–15, and Stearns, *Childhood in World History*. A discussion of the new paradigm of childhood appears in Allison James and Alan Prout, eds, *Constructing and Reconstructing Childhood: Contemporary Issues in the Sociological Study of Childhood* (London: Routledge, 1997), 1–7.
11 For example Elizabeth Buettner, *Empire Families* (Oxford: Oxford University Press, 2004); Karen Vallgarda, *Imperial Childhoods and Christian Mission: Education and Emotions in South India and Denmark* (Basingstoke: Palgrave Macmillan, 2015); David Pomfret, *Youth and Empire: Trans Colonial Childhood in British and French Asia* (Stanford: Stanford University Press, 2015); Vyvyen Brendon, *Children of the Raj* (London: Phoenix, 2006).
12 A 'mission' could be an embassy or a legation, depending of the perceived importance of the post.
13 For example Diana Gittins, *The Family in Question: Changing Households and Familiar Ideologies* (London: MacMillan, 1985); Bill Osgerby, *Youth in Britain since 1945* (London: Blackwell, 1998); Jon Bernardes, *Family Studies: An Introduction* (London: Routledge, 1997).
14 Diana Gittins, 'Disentangling the History of Childhood', *Gender and History* 1:3 (1989): 342–349.
15 A well-known example is James Clifford, 'Travelling Cultures', in *Cultural Studies*, ed. Lawrence Grossberg et al. (Oxford: Routledge, 1992), 96–117.
16 Alison Blunt, *Domicile and Diaspora: Anglo-Indian Women and the Spatial Politics of Home* (Oxford: Blackwell, 2005) and Alison Blunt and Robyn Dowling, *Home* (London: Routledge, 2006).
17 Iain Chambers, *Migrancy, Culture, Identity* (London: Routledge, 1994).

18 Anne Coles and Anne-Meike Fechter, *Gender and Family among Transnational Professionals* (London: Routledge, 2008). Coles and Fechter discuss their decision to use the term 'transnational professional' in their introduction (pp. 1–20). They find 'migrant' problematic owing to its generic qualities but also dismiss 'sojourner' because it suggests an anticipated return to the country of origin and does not indicate the high level of mobility experienced by transnational families. 'Expatriate' was not used because it is applied generally to Westerners and possibly, more narrowly, to commercial sector employees. 'Elite migrants' and 'global capitalists' were rejected as too emotive and not in common use by the families who were the subjects of the book. Interestingly 'cosmopolitan' was treated with caution because many of the individuals studied in detail retain their own culture while living overseas. The book was produced by the International Gender Studies Centre at the Department of International Development at the University of Oxford. Its Foreword was written by anthropologists Hilary Callan and Shirley Ardener.

19 Fiona Moore, 'The German School in London, UK: Fostering the Next Generation of National Cosmopolitans', in Coles and Fechter, *Gender and Family*, 85–103.

20 Roger Goodman, *Japan's 'International Youth': The Emergence of a New Class of Schoolchildren* (Oxford: Clarendon Press, 1990). It is interesting to note that Goodman's study was published in 1990, at the beginning of the decade in which a number of the works on migrancy and identity discussed above were written. This perhaps illustrates a growing consciousness of and interest in the effects of constant travel on the people who experience it.

21 Gittins, 'Disentangling the History of Childhood', 342.

22 On wives, see, for example, Katie Hickman, *Daughters of Britannia: The Life and Times of Diplomatic Wives* (London: Flamingo, 1999); Beryl Smedley, *Partners in Diplomacy* (Ferring, West Sussex: Harley Press, 1990). Hilary Callan, 'The Premiss of Dedication: Notes towards an Ethnography of Diplomatic Wives', in *Perceiving Women*, ed. Shirley Ardener (London: Malaby Press, 1975), and Helen McCarthy, *Women of the World: The Rise of the Female Diplomat* (London: Bloomsbury, 2014), present an alternative point of view.

23 For accounts of post-war family life in Britain see Claire Langhamer, 'Meanings of Home in Post-war Britain', *Journal of Contemporary History* 40:2 (2005): 342–361; Mathew Thomson, *Lost Freedom. The Landscape of the Child and the Post-war Settlement* (Oxford: Oxford University Press, 2013) and Laura King, *Family Men* (Oxford, Oxford University Press, 2015).

24 Brian Harrison, *Seeking a Role: British Society 1951–1970* (Oxford: Clarendon Press, 2009), 195.

25 See Denise Riley, *War in the Nursery* (London: Virago, 1983) 8; Thomson, *Lost Freedom*; Michal Shapira, *The War Inside: Psychoanalysis, Total War and the Making of the Democratic Self in Post-war Britain* (New York: Cambridge University Press, 2013).

26 Brian Harrison, *Finding a Role: The United Kingdom 1970–1990* (Oxford: Clarendon Press 2011), 227.

27 See, for example, Helen McCarthy, 'Women, Marriage and Work in the British Diplomatic Service', *Women's History Review* 23:6 (2014): 853–873.

28 'The century of the child' is often used to refer to developments in thinking about the situation of children during the twentieth century and was taken from title of a book by Swedish educationalist Ellen Key which was translated into English as *The Century of the Child* and published in 1909.

29 See Hugh Cunningham, *The Invention of Childhood* (London: BBC Books, 2006), on children and the welfare state (178); on the tensions between 'behaviourists' and advocates of a more relaxed form of discipline (198–201); on 'investment' in children, emotional and otherwise (215).

30 Thomson, *Lost Freedom*.

31 Ellen Boucher, *Empire's Children: Child Emigration, Welfare and the Decline of the British World 1869–1967* (Cambridge: Cambridge University Press, 2014), see also Philip Bean and Joy Melville, *Lost Children of the Empire* (London: Unwin Hyman, 1989).

32 The Basque children were granted entry into the UK for a limited time, on the condition that they returned to Spain once the war was over. For their history see the website of the Basque Children of '37 Association www.basquechildren.org (accessed 01/08/2016). On the *kindertransports* see Mark Harris and Deborah Oppenheimer, *Into the Arms of Strangers. Stories of the Kindertransport* (London: Bloomsbury, 2001), for oral history accounts of the children's experiences; for a more in-depth analysis of British government policy towards refugee children, the *kindertransports* and the relationship between government and Jewish aid organisations see Judith Baumel-Schwartz, *Never Look Back: The Jewish Refugee Children in Great Britain 1938–1945* (Purdue IN: Purdue University Press, 2012).

33 John MacNicol, 'The Effect of the Evacuation of School Children on Official Attitudes to State Intervention', in *War and Social Change: British Society and the Second World War*, ed. Harold L Smith (Manchester: Manchester University Press, 1986), outlines the short and long term effects of civilian evacuation on British children, often from working class families.

34 Brian Stanley Johnson, ed., *The Evacuees* (London: Victor Gollancz, 1968); discussed in Thomson, *Lost Freedom*, 56–59.

35 Jennifer Mori, *The Culture of Diplomacy* (Manchester: Manchester University Press, 2010), makes the case for the study of a number of areas of diplomatic culture which have only received partial coverage from historians and other observers. Alongside family sex and marriage, Mori also discusses entry into the service, favouritism, gossip and early 'networking'.

36 Hickman, *Daughters of Britannia*, 220–247.

37 See note 25 above.

38 Valerie Cromwell, 'A World Apart: Gentleman Amateurs To Professional Generalists', in *Diplomacy and World Power Studies in British Foreign Policy 1890–1950*, ed. Michael Dockrill and Brian McKercher (Cambridge: Cambridge University Press, 1996); Zara Steiner, 'The Foreign and Commonwealth Office: Resistance and Adaptation to Changing Times', *Contemporary British History* 18:3 (2004): 13–30.

39 Kenneth Weisbrode, 'Vangie Bruce's Diplomatic Salon: A Mid-Twentieth Century Portrait' in *Women, Diplomacy and International Politics Since 1500*, ed. Glenda Sluga and Carolyn James (London: Routledge, 2016), 240–254.

40 Brendon, *Children of the Raj*, 185–212; 241–273.

41 Vyvyen Brendon, *Prep School Children: A Class Apart Over Two Centuries* (London: Continuum, 2009), 7.

42 Royston Lambert with Spencer Millham, *The Hothouse Society. An Exploration of Boarding School through the Boys' and Girls' Own Writing* (Harmondsworth: Penguin, 1974), 11.

43 Paul John Rich, *Chains of Empire: English Public Schools, Masonic Children, Historical Causality and Imperial Clubdom* (London: Regency Press, 1991), deals with what Rich sees as the interconnected rituals between public schools, gentlemen's clubs and imperial rule albeit in an extremely eccentric and eclectic way.

44 Thomas William Bamford, *The Rise of the Public Schools: A Study of Boys' Public Boarding Schools in England and Wales from 1837 to the Present Day* (London: Nelson, 1967), 84.

45 Goffman himself defines a total institution as 'a place of residence and work where a large number of like-situated individuals, cut off from the wider society for an appreciable period of time, together lead an enclosed, formally administered round of life'. Erving Goffman, *Asylums* (Harmondsworth: Penguin, 1968), 11.

46 This anxiety was borne out in December 2013 when the elderly 'child' of a celebrated diplomatic family agreed, with enthusiasm, to be interviewed in the New

Year. However, over the Christmas period he fell ill and his family wrote to apologise that he was unable to take part in the project.

47 An added danger in over-investing time in younger participants is that a clear picture of the later period could be formed without the historical background from older contributors to provide a framework.

48 Leonie Gordon, 'The Shell Ladies Project: Making and Remaking the Home', in Coles and Fechter, *Gender and Family*, discusses her own 'insider-outsider' status as a 'Shell child', 24–25; also Ann Coles, 'Making Multiple Migrations: The Life of British Diplomatic Families Overseas,' in Coles and Fechter, *Gender and Family*, 127.

49 Kirsten Haastrup, 'The Native Voice – and the Anthropological Vision', *Social Anthropology* 1: 173–186 (1993).

50 Gordon, 'The Shell Ladies Project', 21–41.

51 Coles, 'Making Multiple Migrations', 27–147.

52 Cunningham, *Invention of Childhood*, 12.

53 Rich, *Chains of Empire,* 84; see also Judith Okley, *Own or Other Culture* (London: Routledge, 1996), 133–161.

54 Further exploration of the ways in which children and childhood are defined, as well as their 'meaning' in contemporary Britain and the West, can be found in Cunningham, *Invention of Childhood*, 12–16; Fass, *Childhood in the Western World*, 9–12 and 158–171; also Stearns, *Childhood in World History*, 3.

55 Stearns, *Childhood in World History*, 7.

56 I have allowed myself some latitude with these dates and have certainly not dismissed useful evidence as long as it fell within five years of the lower or upper limits.

57 Stearns, *Childhood in World History*, 5. In his inquiry into the challenges facing the historian of childhood, Stearns identifies two interconnected tensions as the uneven nature of sources and cultural differences in the experience of childhood. The experiences of two children of the same age cannot necessarily be treated as in any way similar.

58 Harry Hendrick, *Children, Childhood and English Society 1880–1990* (Cambridge University Press, 1997), 3. Hendrick stresses that 'shrewd remarks' made by middle class investigators are open to interpretation across a range of views and variables.

59 A.O.J. Cockshut, 'Children's Diaries' and Olivia and Alan Bell 'Children's Manuscript Magazines in the Bodleian Library', in *Children and Their Books A Celebration of the Work of Iona and Peter Opie,* ed. Gillian Avery and Julia Briggs (Oxford: Clarendon, 1989), 381–398 and 399–412.

60 *The Double J* is kept by the Dixon family and it is with their kind permission that it forms part of the background of this project. Pierson Dixon (1904–1965) was a career diplomat whose career spanned the periods before and after the Second World War. Dixon's post-war career saw him as UK representative to the Union Nations between 1954–1957(during the Suez crisis) and his final role before retirement was as British Ambassador in Paris (1960–1965).

61 'Views of Children and Parents', *DSWA Newsletter,* Autumn 1980, 29–32; N.J. Morley, 'The International Child' *DSWA Newsletter,* Spring 1981, 37–38.

62 'Calling All Under 16s' *DSWA Newsletter*, Autumn 1982, 23.

63 Heywood, *Childhood in the West*, 6–7.

64 Stearns, *Childhood in World History*, 6–7.

65 The formation of the Diplomatic Service Wives Association is discussed in Chapter 2.

66 Smedley's *Partners in Diplomacy* acts as part history of diplomatic wives and part memoir of Smedley's travels with her husband Harold. Smedley's papers contain the responses to her printed questionnaires, with many lengthy addenda from wives, some of whom were married as early as the 1920s. For the purposes of this book

Smedley's collected material will be identified as originating from Box A (contains questionnaire responses) and Box B (newspaper cuttings).

67 Coles and Fechter, *Gender and Family*, 4. Coles considers why journalism and writing are 'advocated as ideal pursuits' for diplomatic and transnational wives. The work of fashion journalist turned memoirist Brigid Keenan is singled out for mention.

68 Katie Walsh, 'Travelling Together? Work, Intimacy, and Home amongst British Expatriate Couples', in Anne Coles and Anne-Meike Fechter, *Gender and Family among Transnational Professionals* (London: Routledge, 2008), 64.

69 The National Archives, Kew (hereafter TNA) FCO 366/3103.

70 TNA, FCO 53/244; TNA, FO 366/3511; TNA, FCO 77/204 and TNA, FCO 77/205.

71 Report of the Committee on Representational Services Overseas appointed by the Prime Minister under the Chairmanship of Lord Plowden, 1962–63 (Representational Services Overseas), 1963–64 Cmnd. 2276 (hereafter Plowden Report).

72 These volumes are comprised of the minutes of Plowden's Main and Sub Committees and are dated between 1962 and 1963. They were earmarked for disposal before I used them.

73 Rob Perks and Alistair Thompson, eds, *The Oral History Reader* (London: Routledge, 2006), 2.

74 Alistair Thompson, 'The Voice of the Past: Oral History', in Perks and Thompson, *Oral History Reader*, 25–31.

75 Thompson, 'The Voice of the Past', 29.

76 The British Diplomatic Oral History Project (BDOHP) www.chu.cam.ac.uk/archives/collections/bdohp/ contains around 150 interviews with retired diplomats who had achieved high rank.

77 Linda Shopes, 'Oral History and the Study of Communities', in Perks and Thompson, *Oral History Reader*, 261–271.

78 Ann-Laura Stoler and Karen Strassler 'Memory Work in Java. A Cautionary Tale', in Perks and Thompson, *Oral History Reader*, 283–309. Stoler and Strassler caution the oral history practitioner to guard against familiar stories related by former colonials that were perhaps suggested by nostalgia or literary representations of colonial life. Of particular interest to this project is what Stoler and Strassler term the 'popular romance of the beloved and nurturing servant'.

79 Hendrick, *Children, Childhood and English Society*, 5.

Bibliography

Aries, P. *Centuries of Childhood: A Social History of Family Life* (trans. R. Baldick) (London: Jonathan Cape, 1962).

Bamford, T.W. *The Rise of the Public Schools: A Study of Boys' Public Boarding Schools in England and Wales from 1837 to the Present Day* (London: Nelson, 1967).

Baumel-Schwartz, J. *Never Look Back: The Jewish Refugee Children in Great Britain 1938–1945* (Purdue IN: Purdue University Press, 2012).

Bean, P. and Melville, J. *Lost Children of the Empire* (London: Unwin Hyman, 1989).

Bell, O. and Bell, A. 'Children's Manuscript Magazines', in *Children and Their Books: A Celebration of the Work of Iona and Peter Opie*, ed. G. Avery and J. Briggs (Oxford: Clarendon, 1989).

Bernardes, J. *Family Studies: An Introduction* (London: Routledge, 1997).

Blunt, A. *Domicile and Diaspora: Anglo-Indian Women and the Spatial Politics of Home* (Oxford: Blackwell, 2005).

Blunt, A. and Dowling, R. *Home* (London: Routledge, 2006).

Boucher, E. *Empire's Children: Child Emigration, Welfare and the Decline of the British World 1869–1967* (Cambridge: Cambridge University Press, 2014).

Brendon, V. *Children of the Raj* (London: Phoenix, 2006).

Brendon, V. *Prep School Children: A Class Apart Over Two Centuries* (London: Continuum, 2009).

Buettner, E. *Empire Families* (Oxford: Oxford University Press, 2004).

Callan, H. 'The Premiss of Dedication: Notes towards an Ethnography of Diplomats' Wives', in *Perceiving Women*, ed. S. Ardener (London: Malaby Press, 1975).

Chambers, I. *Migrancy, Culture, Identity* (London: Routledge, 1994).

Clifford, J. 'Travelling Cultures', in *Cultural Studies*, ed. L. Grossberg *et al.* (Oxford: Routledge, 1992).

Cockshut, A.O.J. '*Children's Diaries*', in *Children and Their Books: A Celebration of the Work of Iona and Peter Opie*, ed. G. Avery and J. Briggs (Oxford: Clarendon, 1989).

Coles, A. 'Making Multiple Migrations: The Life of British Diplomatic Families Overseas', in *Gender and Family Among Transnational Professionals*, ed. A. Coles and A. Fechter (London: Routledge, 2008).

Coles, A. and Fechter, A. eds. *Gender and Family Among Transnational Professionals* (London: Routledge, 2008).

Cromwell, V. 'A World Apart: Gentleman Amateurs to Professional Generalists', in *Diplomacy and World Power Studies in British Foreign Policy 1890–1950*, ed. M. Dockrill and B. McKercher (Cambridge: Cambridge University Press, 1996).

Cunningham, H. *The Invention of Childhood* (London: BBC Books, 2006).

Diplomatic Service Wives Association (DSWA) Magazine, Autumn 1980; Autumn 1982; Spring 1981.

Fass, P.S. ed. *The Routledge History of Childhood in the Western World* (London: Routledge, 2013).

Faulk, B.J. and Harrison, B. eds. *Punk Rock War Lord: The Life and Work of Joe Strummer* (London: Routledge, 2016).

FCO Files 52/244; 366 /3013; 366/3511;77/204; 77/205 The National Archives, Kew.

Gittins, D. *The Family in Question: Changing Households and Familiar Ideologies* (London: MacMillan, 1985).

Gittins, D. 'Disentangling the History of Childhood', *Gender and History* 1:3 (1989): 342–349.

Goffman, E. *Asylums* (Harmondsworth: Penguin, 1968).

Goodman, R. *Japan's 'International Youth': The Emergence of a New Class of Schoolchildren* (Oxford: Clarendon, 1990).

Gordon, L. 'The Shell Ladies Project: Making and Remaking the Home', in *Gender and Family Among Transnational Professionals*, ed. A. Coles and A. Fechter (London: Routledge, 2008).

Haastrup, K. 'The Native Voice – and the Anthropological Vision', *Social Anthropology* 1 (1993): 173–186.

Harris, M. and Oppenheimer, D. *Into the Arms of Strangers. Stories of the Kindertransport* (London: Bloomsbury, 2001).

Harrison, B. *Seeking a Role: The United Kingdom 1951–1970* (Oxford: Clarendon, 2009).

Harrison, B. *Finding a Role: The United Kingdom 1970–1990* (Oxford: Clarendon, 2011).

Hendrick, H. *Children, Childhood and English Society 1880–1990* (Cambridge: Cambridge University Press, 1997).

Heylin, C. *Babylon's Burning: From Punk to Grunge* (London: Penguin, 2008).

Heywood, C. *A History of Childhood. Children and Childhood in the West from Medieval to Modern Times* (Cambridge: Polity Press, 2001).

Hickman, K. *Daughters of Britannia: The Lives and Times of Diplomatic Wives* (London: Flamingo, 1999).

James, A. and Prout, A. eds. *Constructing and Re-constructing Childhood: Contemporary Issues in the Sociological Study of Childhood* (London: Routledge, 1997).

Johnson, B.S. ed. *The Evacuees* (London: Victor Gollancz, 1968).

Jones, A. 'The Clash Live On!', *UNCUT Magazine*, October2013.

King, L. *Family Men* (Oxford: Oxford University Press, 2015).

Lambert, R. with Millham, S. *The Hothouse Society. An Exploration of Boarding School Life through the Boys' and Girls' Own Writings* (Harmondsworth: Penguin, 1974).

Langhamer, C. 'Meanings of Home in Post-war Britain', *Journal of Contemporary History* 40:2 (2005): 342–361.

McCarthy, H. *Women of the World: The Rise of the Female Diplomat* (London: Bloomsbury, 2014).

McCarthy, H. 'Women, Marriage and Work in the British Diplomatic Service', *Women's History Review* 23:6 (2014): 853–873.

MacNicol, J. 'The Effect of the Evacuation of School Children on Official Attitudes to State Intervention', in *War and Social Change: British Society and the Second World War*, ed. H.L. Smith (Manchester: Manchester University Press, 1986).

MacNicol, J. 'Welfare Wages and the Family', in *In the Name of the Child: Health and Welfare 1880–1940*, ed. R. Cooter (London: Routledge, 1992).

Mori, J. *The Culture of Diplomacy* (Manchester: Manchester University Press, 2010).

Moore, F. 'The German School in London, UK: Fostering the Next Generation of National Cosmopolitans', in *Gender and Family Among Transnational Professionals*, ed. A. Coles and A. Fechter (London: Routledge, 2008).

Okley, J. *Own or Other Culture* (London: Routledge, 1996).

Osgerby, B. *Youth in Britain since 1945* (London: Blackwell, 1998).

Perks, R. and Thompson, A. eds. *The Oral History Reader* (London: Routledge, 2006)

Pomfret, D. *Youth and Empire: Trans Colonial Childhood in British and French Asia* (Stanford: Stanford University Press, 2015).

Report of the Committee on Representational Services Overseas appointed by the Prime Minister under the Chairmanship of Lord Plowden, 1962–63 (Representational Services Overseas), 1963–1964 Cmnd. 2276.

Rich, P.J. *Chains of Empire: English Public Schools, Masonic Children, Historical Causality and Imperial Clubdom* (London: Regency Press, 1991).

Riley, D. *War in the Nursery: Theories of the Child and Mother* (London: Virago, 1983).

Salewicz, C. *Redemption Song: The Definitive Biography of Joe Strummer* (London: Harper Collins, 2007).

Shopes, S. 'Oral History and the Study of Communities', in *The Oral History Reader*, ed. R. Perks and A. Thompson (London: Routledge, 2006).

Smedley, B. *Partners in Diplomacy* (Ferring, West Sussex: Harley Press, 1990).

Stearns, P.N. *Childhood in World History: Themes in World History* (London: Routledge, 2011).

Steiner, Z. 'The Foreign and Commonwealth Office: Resistance and Adaptation to Changing Times', *Contemporary British History* 18:3 (2004): 13–30.

Stoler, A-L. and Strassler, K. 'Memory Work in Java: A Cautionary Tale', in *The Oral History Reader*, ed. R. Perks and A. Thompson (London: Routledge, 2006).

Thomson, M. *Lost Freedom: The Landscape of the Child and the Postwar Settlement* (Oxford: Oxford University Press, 2013).

Thompson, A. 'The Voice of the Past: Oral History', in *The Oral History Reader*, ed. R. Perks and A. Thompson (London: Routledge, 2006).

Tompkins, J.H. 'Joe Strummer's Complex Legacy: 10 Years After the Clash Leader's Death', *Spin Magazine*, December2012.

Vallgarda, K. *Imperial Childhoods and Christian Mission: Education and Emotions in South India and Denmark* (Basingstoke: Palgrave Macmillan, 2015).

Walsh, K. 'Travelling Together? Work, Intimacy, and Home amongst British Expatriate Couples in Dubai', in *Gender and Family Among Transnational Professionals*, ed. A. Coles and A. Fechter (London: Routledge, 2008).

Weisbrode, K. 'Vangie Bruce's Diplomatic Salon: A Mid-Twentieth Century Portrait', in *Women, Diplomacy and International Politics Since 1500*, ed. G. Sluga and C. James (London: Routledge, 2016).

1 1945–1958

Diplomatic Service society after the Second World War

Introduction

That the Diplomatic Service is a family affair has long been part of its rhetoric. Mori writes that 'surrogate fatherhood was conferred' on those eighteenth century envoys who established and equipped their own embassies and it was often the case that their staff were related by blood or via some other family connection.[1] Cromwell also observes that the family concept was never far from the forefront of the Foreign and Commonwealth Office (FCO)'s official self-image. 'It was the enduring concept of the mission abroad as a family, and a particular type of family at that, which was to characterise the service till long after 1919.'[2] Marcus Cheke's 1949 'Guidance ... for a member of His Majesty's Foreign Service on his first appointment to a post abroad' explicitly underlined the patriarchal hierarchy of a British overseas mission: 'The whole Embassy forms a sort of family ... of which His Majesty's Ambassador is *pater familias* ...'[3] The reality (of embassy life in particular), however, was very different. During the post-war period – defined in this chapter as the years between 1945 and 1958 – the Foreign Office administration maintained that 'wives and children had never been their concern'.[4] There was very little financial aid available for Diplomatic Service families and no systems of pastoral support existed at all. Thus, immediately after the Second World War, the Diplomatic Service did not provide a welcoming environment for the families of ordinary middle class meritocrats starting out on a career in government service.

This chapter begins with a discussion of the British Diplomatic Service's far from flexible attitudes towards social distinctions in this era and it demonstrates that before, during and after the Second War World the Service retained fixed ideas about the preferred social pedigree of its members. As will be seen the Diplomatic Service family retained strong links and shared modes of behaviour with the aristocracy and upper class British families and its public image was closely associated with this group. The first section of this chapter examines families in this context and demonstrates the way in which families of lesser social standing could undergo 'the process of acculturation' identified by Cromwell, by adopting the values and attitudes of established diplomatic families. The second section turns to the optimistic social reforms of the post-

DOI: 10.4324/9780429273568-1

war Attlee government and examines how the role of home and family became central to the British post-war reconstruction. It is a compelling aspect of this study that it was at this point that the experiences of Foreign Office children began to deviate significantly from those of their counterparts at 'home' in the UK: 'home' as a concept being something that children who led mobile lives found difficult to grasp and comprehend. The final section introduces the reader to the Diplomatic Service culture of family separation, still identified as late as 2004 as 'one of the worst aspects of diplomatic life'.[5] It analyses the reasons for this practice and examines the way that Diplomatic Service practice intersected with that of similar groups, for example colonial administrators and army and missionary children. It details the experiences of some of the children who experienced separation and looks at the systems of care provided by extended families and commercial agencies in an international environment.

Diplomatic Service families at the time of the Second World War

More than once during the early twentieth century attempts were made to recruit civil servants into the Foreign Office from more varied social backgrounds. The first, the Royal Commission on the Civil Service, pre-dated the First World War and many of its recommendations were never implemented because of it. The second attempt at change, *Proposals for the reform of the Foreign Service Presented by the Secretary of State for Foreign Affairs to Parliament by Command of his Majesty*, published by HMSO in 1943, was testament to Eden's dedication to the ideal of a more equitable post-war Foreign Service. Nonetheless, at the close of World War Two the Diplomatic Service was still dominated by aristocratic and socially distinguished families. Cannadine noted that 'Between 1873 and 1945, eleven men held the post of Permanent Under Secretary: nine were peers, close relatives of peers, or bona fide landed gentry; only two came from the middle classes'.[6] Diplomacy was an exclusive world, jealously guarded by its inhabitants. The urge to restrict this environment to socially acceptable candidates was illustrated shortly before the Second World War began, when a Departmental Committee was set up to discuss the amalgamation of the Diplomatic Service and Consular Service. The latter had always been known informally as 'The Cinderella Service' owing to its less splendid aspects and there were many meticulously observed differences: for example, diplomats were entitled to gold stitching on their official uniforms, where consuls had to make do with silver thread. Members of the Diplomatic Service vigorously opposed the suggested merger: in a private letter to the Permanent Under Secretary, Sir Alexander Cadogan, in January 1939, diplomat Sir Hughe Knatchbull-Hugessen made the sentiments of the Diplomatic Service towards the Consular Service and, by extension, the amalgamation of the two Services, clear:

Though we should be far from suggesting that personality, 'address', and *savoir-faire* are not of great importance in the Consular Service it is in the Diplomatic Service that these rather intangible qualities are most essential. A diplomatic officer must be prepared on all occasions to represent the most representative orders of his own countrymen. He must be able to deal as an equal with foreign colleagues, Cabinet Ministers, Prime Ministers and Heads of State; to hold his own with Sovereigns and other royalties and to fraternise with the governing class in no matter what country ... there is a danger that the Service may suffer by the admission of candidates who lack the qualities which we have tried to describe.[7]

In this letter we also see that Knatchbull-Hugessen emphasised the importance of the diplomatic wife as 'Chefesse', observing that:

In many posts the part played by the wife is fully as important as that of the husband, and in all an immense influence for or against British prestige is exercised by the wife ... We feel that in selecting officers for service abroad, whether in an amalgamated service or not, great attention should be paid to this consideration.

Lower class diplomats would presumably be married to lower class wives. Knatchbull-Hugessen was anxious to stress that the arguments given in his letter were unsuitable for inclusion in an official report; 'difficult and delicate' is the way that he put it. This was an early illustration of the ambiguity that surrounded the role of Diplomatic Service dependents, and wives in particular, during the twentieth century. According to the official Foreign Office administration, dependents were not important and yet the 'intangible qualities' that they possessed were of the utmost importance, and the lack of them precluded advancement within the service.

The 1943 report was dedicated to reforming these kinds of distinctions and the attitudes that upheld them within the Foreign Service. Its initial proposals criticised the Diplomatic Service, appearing to provide a direct response to the points made by Knatchbull-Hugessen: 'the view has been expressed that it has been recruited from too small a circle, that it tends to represent the interests of certain sections of the nation, than those of the country as a whole ...'[8] However, a wartime recruitment freeze meant that 'The new system of recruitment and training ... will not ... be felt for some years after its introduction'.[9] For the moment, then, during the immediate post-war period, diplomats and their families continued to enjoy the 'uniquely genteel tone' that membership of the Diplomatic Service bestowed.[10]

Both the sender and recipient of the letter that discouraged amalgamation within the Foreign Office, Knatchbull-Hugessen and Cadogan, were firmly rooted in the British upper class. The former came from a long line of baronets, while the latter was son of the 5th Earl of Cadogan. Similarly, other Diplomatic Service families in this period were already well-known figures in British high society. Duff Cooper, first British ambassador to Paris after the German occupation, was married to the celebrated aristocratic beauty Diana Cooper (born Diana Manners, daughter of the Duke of Rutland), who was thought to have been the model for Lady Leone, the beautiful wife of the outgoing ambassador in Nancy Mitford's novel *Don't Tell Alfred*, based in the British Embassy in Paris. Edward Wood, Lord Halifax, 'a quintessential grandee', was another aristocratic figure who typified the glamorous image of the mid-twentieth century Foreign Office. Halifax was described as a 'calm, rational man of immense personal prestige and *gravitas* ...'[11] In a career that effectively traversed the divide between the institutions of empire and diplomacy, he was Viceroy of India between 1926 and 1931, served as Foreign Secretary from 1938 to 1940 and then took on the role of British Ambassador to the United States during World War Two. Grand as he was, Halifax enjoyed his large and visible family which 'lent the [Viceregal] Court a relaxed atmosphere'. His choice of spouse was, according to his biographer, crucial to his success in these varied posts: 'Lady Dorothy fulfilled a vital role in complementing her husband's image of a sound family man who could be trusted to do the decent thing.'[12] The establishment position of figures like Halifax and Lord Curzon, a similarly patrician figure twenty-two years Halifax's senior, who also held the offices of Viceroy and Foreign Secretary, no doubt contributed to the incorrect but widely held perception that imperial administration – especially that of India – and diplomacy were interchangeable. (Although there were many similarities: the Foreign Office shared a building with the India and Colonial Offices and many well-known diplomats of the later twentieth century were 're-treads', former colonial officials who joined the Foreign Office as the decolonisation process saw overseas administrations gradually disbanded.)

Sir David Kelly, a diplomat between 1919 and 1952, went to great pains to uphold the Diplomatic Service's exclusive image. In his memoir *The Ruling Few,* Kelly famously aligned the early twentieth century Diplomatic Service with the 'King's Household ... not really part of the Civil Service at all'.[13] In common with other influential twentieth century diplomats (the Gore-Booths, for example), Kelly came from the Anglo-Irish gentry and his second wife Rene Marie Nole Ghislaine de Vaux, known as Marie-Noele, was a Belgian aristocrat who became a celebrated hostess, both within diplomatic circles and in her later years. Like her husband, Marie-Noele Kelly was very much attached to the ideal of diplomatic life as leisured, instinctive and with time for other suitable pursuits; she also felt that her observations were valuable enough to command a readership. The volume of memoirs, *Dawn to Dusk,* which Marie-Noele Kelly wrote in 1960, eight years after David Kelly's death, was a vehicle for her expansive and didactic personality and it adds a great deal to our understanding of what life was like for families like this before, during and after the Second World War. The Kelly family's aristocratic family values provide an indication of their striking differences and similarities to contemporary family life.

Kelly's memoirs were addressed to her grandson Dominic, and suggested that her life contained many circumstances of note: 'My grandchildren, when they read this, may wonder if I met any of the men who made history.'[14] This was no exaggeration. Kelly's family, in Belgium, suffered greatly during both world wars; diplomatic life introduced her to world travel and brought her into contact with many of the twentieth century's great statesmen: in *Dawn to Dusk* she described meeting Neville Chamberlain, Winston Churchill, Clement Attlee and Ernest Bevin. The Kellys' austere last posting, in Soviet Russia, was a noteworthy contrast to the 'parties, balls and huge dinners' that had been a feature of the past.[15] In the introduction to her book, Kelly admitted that she had always valued family life 'more than any other' and it is clear from the way she wrote about herself and her family that the purpose of her memoirs was to stress dynastic continuity and reputation. Kelly's description of her own upbringing demonstrates a heightened awareness of social discipline, responsibility and status, not only as applied to her family with regard to others but between its generations. She received 'social training in manners and outlook which we received in common with all the young of our class'.[16] Nuances of family pedigree and belonging were also very important; 'even the most distant and seldom-met had to be addressed as "my cousin" or "my uncle"'.[17] The life was excessively formal: 'It was unthinkable in my time for children to be "amused" as a system …' Kelly wrote. These aristocratic children were 'loved by their own parents in their own way'.[18] The exaggerated sense of observance and duty taught to Marie-Noele Kelly during her own childhood and youth corresponded well very with that demanded by the Diplomatic Service from its families. It is typical of aristocratic European society in the early twentieth century and reminiscent of a phrase in the letter from Knatchbull-Hugessen that began this section: 'A diplomatic officer must be prepared on all occasions to represent the most representative orders of his old countrymen.'[19] Her memoirs indicate plainly that Marie-Noele Kelly was not a cold or neglectful parent but her approach to what we would now term 'parenting' was very different to many of the models conspicuous today: informed by public service and its inherent values of duty and sacrifice, in common with the Diplomatic Service itself.[20] Thus she was able to write without fear of criticism: 'I had left our eldest son, Bernard, at school at Downside. I knew I should not see him for a year or two …'[21]

The children of the type of families described above, of men who held high government office, were accustomed to the rituals and duties of their class; 'many a diplomat's memoirs wax lyrical on the elaborate court the country house visits, and the opportunities for travel …' wrote Cannadine.[22] Rituals like official photographs could include all members of the family, like the one of the Kelly family pictured in 1955 which is in the National Portrait Gallery collection in London. Children of well-known diplomatic figures could find themselves subject to press attention: like John Julius Cooper, the son of Duff and Diana Cooper who encountered the press on his arrival as an evacuee in the US in 1940. He wrote in a letter to his mother: 'as we were queueing up to have our passports etc., examined, lots of reporters came on board. We kept them off for about twenty minutes but they knew I was there and they were so persevering ...'[23] The children of less prominent families could also find themselves in the spotlight. Olivia Tate*, a diplomat's child born in 1946, recalled:

> Sometimes you get dragged into things perhaps that you wouldn't otherwise ... I was doing singing at school and the same year I came out for Christmas and my mother said you're doing a solo in the church choir [laugh]. But ... it was not BEING BRITISH, it was you know [lowers voice to suggest modesty and reserve] being *British*...[24]

A look at the domestic life of Pierson Dixon, a diplomat who held a number of senior posts during and after the Second World War, culminating as Ambassador to France between 1960 and 1965, provides a useful contrast to the type of family described above. Dixon was described as 'one of the ablest diplomatists' by Anthony Eden, for whom he worked as Private Secretary, yet, unlike the breed of diplomats suggested by Knatchbull-Hugessen, he grew up in 'one of a string of Victorian brick built houses' in Bedford and is quoted as having said that 'We were very badly off and under-nourished and cold'.[25] An inveterate intellectual, said to retain a 'donnish and detached manner', Dixon was a scholarship boy who had to choose between a career as a Cambridge academic and the Diplomatic Service which he joined in 1929.[26] Although, according to Ellison:

> early on, he did abandon the 'more discursive academic idiom' and developed what his first FO superior described as the qualities prized in the bureaucrat: 'a clear and logical mind with a capacity for rapidly getting at the basic essential of a problem, combined ... with a keen appreciation of human values'.[27]

The Dixons' home life in Surrey was documented in a manuscript magazine, *The Double J*, by the two oldest children John and Jennifer between 1941 and 1948. In opposition to the world inhabited by the Kelly and Cooper families, *The Double J* reveals the Dixons as disorganised, humorous and informal: far closer to a modern idea of family. Pierson Dixon is clearly beloved by his children, referred to throughout as 'Daddy', 'Mr Dixon' or 'The Great Man'. But Dixon had no pretensions to grandeur: one entry for April 1941 notes that 'the Great Man' who had recently returned from an exhaustive tour of Europe with Eden and Churchill has been on holiday and that 'much of the time was spent sowing spring seeds'.[28] When Dixon's wife Ismene gave birth to their third child Corinna in August 1941 John and Jennifer recorded that he visited her and the baby in the nursing home every night. And when Corinna Dixon learned to walk Dixon made her a stair gate which he fitted himself, despite the seniority and demands of his role in the wartime Foreign Office.

The Dixon children were very confident in their roles within the extended family; they clearly found the adult members accessible and respectful of the opinions they were encouraged to have. They regularly poked fun at the grown-ups through *The Double J*'s 'Humorous Corner', often at the expense of their grandmother, the timid Mrs Atchley. They recorded this characteristic exchange, which manages to capture Mrs Atchley's outmoded opinions and their Aunt Peggy's far more modern outlook in August 1941:

MRS ATCHLEY: (startled, half entering half backing out of the dining room door): Oh, I thought I saw a man's legs.
MISS MARGARET DIXON: It's all right, Mrs Atchley, don't be alarmed, but *do* admire my beautiful new corduroy slacks.[29]

The Dixon children had a good knowledge of the political situation, the progress of the war and a distinct pride in their father: they specifically noted that he had represented Anthony Eden, the Foreign Secretary, at meetings. John showed a sophisticated curiosity about the situation in wartime Europe and often used *The Double J* to record his comments on it. This was largely through the discussion and representation of maps. Although the family was represented formally as Dixon's career developed, for example in official photographs, *The Double J* shows that they were able to balance the formal demands of Diplomatic Service membership, making sense of their place within it, by celebrating informal family ties.

However, despite the fact that Dixon's own upbringing was not privileged, by 1947 he was visiting Eton College to see John who was a student there. John was able to make use of his father's elevated position to invite him and Foreign Secretary Ernest Bevin to address the Political Society.[30] This example of social mobility suggests that the process of 'social acculturation', identified by Cromwell as necessary 'if … new recruits were to be assimilated into the life-style in the diplomatic service', was as applicable to families as it was to individuals.[31] The Dixons' social trajectory – that of a family headed by a highly professional, intelligent and successful man from a less illustrious background that quickly found itself in an elevated social position – became an increasingly characteristic pattern in Foreign Office family life as the twentieth century progressed.

Notwithstanding the impressive social pedigree of many senior diplomats – or the urge felt by newcomers to emulate them – the social upheavals of the twentieth century, including the two world wars, prompted the Diplomatic Service to continue with its scheme to democratise its recruitment policies. It was noted above that the first set of recommendations, made by the Mac-Donnell Royal Commission in 1914, were never successfully implemented, although they did result in the discontinuation of 'a minimum income of £400 a year', formerly a requirement for Foreign Service candidates, in favour of a salary paid to new recruits. This was so that the Foreign Office could widen its field of interest to include men from diverse social backgrounds, although, according to McCarthy, 'the Diplomatic Service retained its aura of exclusivity'.[32] In January 1943, however, a White Paper, *Proposals for the Reform of the*

Foreign Service, was launched with the aim to 're-equip the Foreign Service to meet modern conditions' and aired concerns that the Diplomatic Service was 'recruited from too small a circle'.[33] The White Paper criticised its archaic structure, stating that 'the conditions that the Diplomatic Service originally grew up to meet, no longer exist unchanged in modern international affairs'.[34] Recommendations hinged primarily on educating recruits. Observations that many diplomats had too narrow an understanding of contemporary domestic and international affairs led to recruits spending time in other Whitehall departments; in some cases there would be an emphasis on language tuition; and the preliminary Foreign Office selection board with its emphasis on private means and social contacts was abolished. Finally brought into effect was the amalgamation of the Diplomatic and Consular Services (alongside the Commercial Diplomatic Service), the union discreetly opposed by the Diplomatic Service for 'difficult and dangerous' reasons in 1939.

Family life in Britain after World War Two

After the horror and upheaval of the Second World War, the ideal of home and family became central to Britain's programme of reconstruction, while children became symbolic of its hopes for a new era of peace. Heywood thought that:

> It was the personal experience of the whole population of Great Britain, whose lives were directly affected and disrupted by the conditions of war which rediscovered for the nation the value of family … the family had been undermined by powerful social and economic factors and this loss of status was reflected in its smaller size and the lack of support given to it in social legislation.[35]

Scholars of twentieth century childhood have also argued persuasively that the changes in British post-war family life were an intrinsic part, in the words of Thomson, of the 'social, economic and ideological settlement forged out of the experience of war'.[36] Shapira goes further to suggest that for the first time 'The emotional lives of children as future citizens became a matter for public and official concern and an issue that required expert knowledge and guidance …'.[37] High hopes were invested in children as the individual legislation that formed Attlee's Welfare State was rolled out and the increase in Welfare Services was accompanied, within the home, by a relaxation in parental discipline. The child – rather than the father as in the past – became central to the family: the hopes of parents becoming 'inseparable from the happiness of and success of their children'.[38]

In 1946 the Curtis Committee was formed by the government to investigate 'ways of providing for children deprived of a normal home life' and the irony of this statement when applied to Diplomatic Service family practice is striking, as will be seen. The Committee heard from a number of expert witnesses, among them Donald Winnicott, a psychoanalyst who, along with his Tavistock Clinic colleague John Bowlby – the architect of attachment theory – 'advocated new ways of looking at the relationship between mother and child and the consequences of an interruption in this relationship'.[39] Winnicott 'assigned psychological value to the concept of "home" – already seen as centre of life and comfort for many middle class families and, increasingly, more working-class families'.[40] 'Home' writes Thomson, 'was idealised as all that children needed.'[41] This sentiment was echoed by other family members, war weary and reunited after long periods of separation. Contributors to the 1942 Mass Observation Survey discussed in Langhamer's *The Meanings of Home in Postwar Britain* clearly viewed home and family as synonymous. One female correspondent is quoted as saying: 'A happy home and family life is the bulwark of a Nation.' A sentiment which, Langhamer observed, 'might indeed be taken as *the* blueprint for postwar reconstruction in Britain'.[42]

The Curtis Committee viewed children 'as beings who could be damaged by the lack of individualised care and affection'.[43] Its recommendations favoured a progressive approach to disadvantaged children: for instance by placing practices centred in home and family such as fostering and adoption above institutional care. Most notably the Committee's findings led to the standardisation of childcare policy across local government authorities and ensured that, two years later, these suggestions were implanted in the 1948 Children Act.

How did developments affecting British families during the twentieth century – 'the century of the child' – influence the lives of diplomats' children?[44] The tension between child-centred methods of care (which had rapidly entered the public consciousness as the 'correct' way) and the long-established patterns of distance and separation achieved via nannies, au pairs and more significantly boarding schools, and favoured by the British upper class and British colonial administrators, is one of this book's central and recurring themes. For it was at this point in the century that the behaviours, practices and values of Diplomatic Service families began substantially to diverge from those of the British families based permanently in the UK that they were supposed to represent. Peter Boon (born 1942), the child of former colonials who moved on to the Foreign Office after Indian Independence in 1947, recalled 'it was always in the back of our minds that we were a different kind of family'.[45]

The course of this deviation comprised elements that were both superficial and profound. Material comparisons were inevitable. When David and Marie-Noele Kelly arrived in Argentina during wartime, in 1942, they found that 'the values in Buenos Aires were essentially material; clothes, jewels … food, luxury, travel and, among men, the price of cattle were the basic topics of conversation'. Marie-Noele noted that the structure of Argentinian society 'meant glittering nights with luxurious dinners … Back at the Embassy I would rush upstairs afterwards to read the Reuters telegrams … to know what was happening before seeing it in the papers the next day'.[46] Food and housing were two specific areas where post-war Britain invited an instant and unfavourable comparison with life at post. Denise Holt (born 1949) noted that, for many years after the war, the material gap between Britain and other nations had widened. She recalled:

> compared to peers in post-war Britain (grey and poor), we lived a charmed life, mainly in nice houses or flats, in interesting places, with domestic staff. Embassies in those days were huge and very sociable, so there were constant parties, picnics and expeditions ... my parents definitely had a better standard of living in the service than would have been the case at home.[47]

Peter Boon, whose parents were posted to New York in the mid-1950s, was thrilled with what he found, referring to the America he visited as 'a very sophisticated society'. He described a range of experiences that very few British teenagers of his generation would have had:

> Going to these amazing cinemas. I remember seeing *Ben Hur* ... and then the Radio City music hall was a super place with lovely shows ... Everything to do with the USA was exotic and exciting and just totally ... totally different. Yes ... the buildings were bigger and the cars were faster and the cars were bigger ...[48]

Even at potentially inhospitable posts, for example in Soviet Moscow, foreign diplomats had access to the type of food that was unknown or prohibitively expensive while rationing was still in force in Britain. Olivia Tate★, whose family was posted to Moscow between 1952 and 1954, remembered that her parents had returned home from the ballet: 'and my father said he could do with a snack and my mother opened the fridge and said "Well I'm sorry darling there's only cold pheasant or caviar!"'[49]

Not only did diplomatic families posted overseas during this period enjoy commodious living spaces in prestigious areas during a period of housing crisis in the UK, but they also benefited from domestic servants, a resource the British middle class had seen diminish throughout the first half of the twentieth century, before vanishing into the war effort. 'The Census shows that the number of women in residential domestic service declined from over two million in 1931 to 750,000 in 1951 and 200,000 in 1961.'[50] Marie-Noele Kelly was among many diplomatic spouses forced to complain about the contrast between overseas and 'home': 'From the staff-studded glories of the ambassadorial purple, I turned in a day to the temperamental hisses of my gas cooker.'[51]

Servants were an integral part of the diplomatic household; often they worked in the same house for successive families and were a great source of knowledge and companionship. Some relationships between employer and employee outlasted the length of a posting. So much so that Peter Boon remembered meeting his father's former servant at Karachi airport in 1953 on the way to visit his parents at post in Singapore: 'an old retainer of my father's from India, a servant who had at partition – he was a Muslim – had emigrated with his family to Karachi; he came to the airport and we talked at the airport hotel.'[52]

Obviously servants could be an invaluable source of practical help. This was particularly useful at postings where the political regime could make life difficult. When her husband David was appointed ambassador to Moscow in 1949, Marie-Noele Kelly accompanied him. She wrote: 'The milk must be collected by a Russian servant, who queued for hours every morning and sometimes came back with none.'[53] But domestic help – especially in 'iron curtain' posts like Moscow – could have another side. A number of examples exist of diplomatic families living with spies who posed as servants and reported to communist state regimes during the Cold War era. Olivia Tate★, in Moscow with her family in the early 1950s, remembered that the family home came with a housekeeper called Luba:

> We had a lovely lady called Luba who cooked, 'did' for us and also spied [sic] for us but she and my mother worked it all out so that my mother kept a good cook and Luba had something to report but nothing too drastic.[54]

Hickman has commented on this phenomenon, noting that 'Most diplomatic families, accepting this to be the case, took it in their stride …'[55] and McCarthy reported similar findings: when Mary Galbraith, a young female diplomat who joined the FCO in 1951, was sent on a first posting to Budapest she was subject to the attentions of her maid, Ada, also a spy. 'I didn't mind at all …' Galbraith was quoted as saying. 'This was very much the way it worked.'[56]

Other crucial differences between the experiences of children from Foreign Office families and UK children were less tangible, linked to expectations of diplomatic children to play a role and to the demands of life in the Diplomatic Service which quickly became normalised even for the very young. Foreign Office children were expected to be responsible, resilient and self-reliant, not simply because their parents felt that these were admirable competences for the future but because the Diplomatic Service dictated it should be so; as Coles has observed, the diplomatic family was 'defined, characterised and regulated' by the Foreign Office.[57] Unlike the restructured, child-centred, British family at home in the UK, it was the Service that was central to the diplomatic family: not its children. The children of diplomats relate many anecdotes that saw them involved with entertaining at official functions. Peter Boon recalled arriving in New York for Christmas holidays in the late 1950s; the phrase he used (about two-thirds of the way through this excerpt) '*you could be thrown into it*' is indicative.

> I was only just arrived and with jet lag and so on and my parents said to me 'Look there's the daughter of the *Daily Telegraph* correspondent in New York and … her partner can't take her to the Christmas end of term ball. Would you accompany her?' So I did. And that was quite, quite an

experience, I don't remember much of it but I was half asleep poor girl I think ... you could be thrown into it ... it's interesting about the question of Britishness and identity; I perhaps saw myself as you know, representing Britain at this American High School event.[58]

Representation – most often expressed through formal entertaining – has always been a crucial part of diplomatic life. Mori's observation about eighteenth century diplomatic practice is just as applicable to this period: 'Diplomacy has always been a lifestyle requiring its disciples to abandon customary distinctions between public and private life ...'[59] Meanwhile McCarthy has noted that the 'separate spheres' ideology of gender-based private and public influence which gained ground during the nineteenth century 'seems especially unhelpful in the case of women of rank and wealth who ... were the most likely to be moving in diplomatic circles ...'[60] We have already seen how spies doubled as servants in western diplomatic households during the Cold War period; another of the many areas in which the diplomatic family home diverged from middle class practice in Britain was the terrific intrusion on home life caused by the requirement to entertain guests at very little notice and when, as sometimes happened, political and public life encroached on family routines. In Beirut during the 1958 Lebanon civil war one contributor remembered lying on the floor at the back of the car as her father drove her and her sister to school. Daily life could also come to a sudden and disorienting stop: Peter Boon's parents were evacuated from Egypt in 1956 during the Suez crisis. The 'advancing home-centredness' described by Harrison as characterising post-war Britain, accompanied by the respectable notions of 'keeping yourself to yourself', were not possible for diplomatic families serving overseas, even if they had been the inclination of the individual family.[61]

However, it was not only less glamorous dwelling places and insufficient food that caused diplomatic families, especially their children, to feel uncomfortable and 'not at home' when they arrived in Britain after time overseas. Buettner's description of late imperial children can be applied just as well to children of the Diplomatic Service in the mid-twentieth century: 'Home meant their national homeland, Britain ... somewhere "of which they simultaneously claimed to be a part yet of which they had limited or no direct knowledge".'[62] If we keep in mind Hall on migration, 'Migration is a one way trip. There is no "home" to go back to', how are individuals who make, after Coles, 'multiple migrations' able to make sense of 'home' – in this sense a 'homeland' – as a concept?[63] Denise Holt recalled that when she and her family were overseas they 'talked endlessly of going home, although [when I was] a child we didn't actually have a home in the UK'.[64] This suggests an idealised picture of 'home' rather like the one shared by colonial administrators and early immigrants, and one that could quickly turn to disenchantment. The country diplomatic children called 'home' and that was understood to be their 'homeland', the country of which they, according to their documents, were officially a part, was one that was largely unknown to them. 'Ah, that word – "home."

It was always a complex, contested issue for me,' one correspondent recalled.[65] The enduring imperial notion that exiled Britons would instinctively recognise that they had arrived 'home' was a cherished construct rather than a reality, sustained by conversation or letter. In a discussion of the correspondence of one Anglo-Indian family, the Talbots, Buettner observed that 'their dream of home – with its double connotation of familial domestic sphere and national homeland – had to remain no more than a fantasy, a "castle in the air"'.[66]

As the time period covered by this book progressed, greater encouragement to self-reflection – encouraged by a counselling boom in the 1950s and 1960s discussed at length by Shapira – among young people saw them examining issues such as identity in greater detail.[67] Although self-examination was less common in the immediate post-war years, children from Diplomatic Service families struggled with issues of identity – especially those of national identity – and belonging, alongside other 'service' counterparts. These doubts were perhaps further complicated by the decline in Britain's status as a world power after the Second World War and the accelerating pace of decolonisation. British colonies were administered by a different government service, that operated along different lines, but their existence as symbols of Britain's far reaching influence and authority made a great contribution to the wider 'service' consciousness. Denise Holt summed up this ambivalence well:

> We were never in any doubt of our Britishness: quite the opposite. We lived in a British bubble. For example we sailed to Japan on board a military family ship, with most of the mothers heading for Singapore and Hong Kong. At post, most of our friends and social life derived from the British Embassy ... Among my earliest memories are the Coronation, and the annual QBP's [Queens Birthday Parties] ... We were very proud of being British in the post-war world (although Suez was a blow), and very conscious of the global reach of the UK – the Empire was however shrinking and we were also aware of that ...[68]

The culture of separation

During the period 1945 to 1958, many well-established schools, working to European curricula, existed in cities with large western populations, for example in Buenos Aires and Shanghai. In 1924 and 1947 international schools were founded in Geneva and New York by employees of the League of Nations and the United Nations respectively, both as positive responses in the wake of devastating global conflict. These schools were set up to represent the interests of more than one nation, to reflect the remarkable term 'worldminded', originated in 1957, which is defined as a state of mind possessed by those who 'favour a world-view of the problems of humanity, whose primary reference group is mankind rather than Americans, British, Chinese etc'.[69]

Although many diplomatic children briefly attended an international or local school when they were very young and at post with their parents, in the main, international schools or schools located 'in-country' were not a long-term option for British diplomatic families. If there really was no option, diplomatic parents favoured European schools. In 1948 when Peter Boon was six, his father was posted to the Belgian Congo. There Peter attended the local Belgian school: 'I already had some knowledge of French but I had to learn French in order to be a pupil at the school because everything was in French … that was about eighteen months there.'[70] Olivia Tate★ travelled with her family to Moscow in 1952 where she attended the Anglo American School. Despite the similarities in language, American schools were widely disliked and avoided by British diplomats, who thought they were undisciplined and the academic standards low. Note that Tate★ obliquely expresses this point of view in the excerpt below:

> I was six so I'd done two or three terms in an English Primary School which I remember and then I went to the Anglo American School Moscow but of course that was American education and they don't start as early as the Brits … I was fortunate. I could read before I went to school and I was reading well before I went to school you know Winnie the Pooh was no problem. And I rather relaxed for a couple of years.[71]

The older the child, however, the less likely they were to attend an international school or any other type of school at post. This section will explore the origins of family separation as a way of life. This culture of family separation endured throughout the period covered by this book and was one of the foremost characteristics of Diplomatic Service family life, the most marked difference between diplomats' children and their peers in the UK during the period covered in this chapter.

Family separation had always existed at the interstices of two minority British cultural traditions. The first was that of the comparatively small nexus of groups that serviced Britain's Empire or other of its interests overseas. Children of the Diplomatic Service might have received little scholarly attention, but satisfactory comparisons exist between them and the children of colonial administrators, whose history has been more adequately covered, by Buettner, for example, and Pomfret.[72] The second consisted of the children – generally the sons – of the upper and upper middle classes who had always been sent away to be educated, earlier in the households of other families and later to public schools. The groups came together in a third configuration during the nineteenth century expansion of public schools expressly developed to prepare middle class boys for imperial administration.

As the nineteenth century wore on, the prototype of the united diplomatic couple working in tandem, like those colonial couples described in the work of Procida, was established.[73] As both members of a partnership became more involved in life at post whether as colonials or diplomats, the practice of sending children to school or suitable foster care in Britain developed as standard procedure. Contributors to this book – which concentrates specifically on

Diplomatic Service practice – have described the pattern of small children travelling to post with their parents before being sent 'home' to boarding school
for secondary education as the dominant convention right up until the late
1980s.

The reasons for these often long-range and lengthy separations originated in
a fear of both literal and figurative contamination. Tropical diseases, of course,
were a very real threat to young children, especially in far flung posts where
medical care was limited. Despite the commodious housing and retinues of
servants discussed above, unfamiliar environments could be hostile. Poor infra-
structure, especially sanitation, was treacherous and gardens could provide a
habitat for toxic plants and venomous insects. According to nineteenth century
medical orthodoxy, the tropics would render the European child 'slight, weedy
and delicate' or 'weak and weedy, deficient in energy and lacking in strength';
this often quoted truism, still referenced in the 1980s, is cited by both Pomfret
and Buettner.[74] Additionally, the belief persisted that a tropical climate was
unsuitable for European children because it promoted the early onset of
puberty and sexual interest. British MP Richard Crossman, serving on an
Anglo American Committee of Inquiry in Palestine (then under British
Mandate) in 1946, recalled the comments of a British official's wife, who
summed up the complex mixture of racial and social preconceptions along
with received wisdom on health and welfare when she told him that,
although schools in Jerusalem were 'quite good ...' children who stayed in
Palestine to attend them would 'mix with strange people, mature too rapidly
and get the wrong ideas ...'[75]

Plus as the nineteenth and twentieth centuries advanced British colonials
became wary of diluting their power and influence and increasingly careful to
preserve their superior racial status. This was another reason they preferred their
children to be educated in Britain.[76] According to Buettner, even British
families who had long associations with India were reluctant to send their
children to school there, even with the threat of war in Europe: 'Long standing
objections to these institutions and their mainly "country born" pupils were so
deeply ensconced that many families did not consider them plausible options
even when international conditions otherwise made them logical temporary
choices.'[77] Buettner is frank about the material rewards and enhanced social
standing available to the British in late imperial India. She also recognises that
these families were characterised by 'a discourse of family sacrifice'.[78]

This sense of sacrifice became a crucial ideal of 'service life' demanding high
levels of emotional resilience and control from both children and parents.
Diplomatic Service children were required, early on, to develop strategies to
help them to deal with the sudden and drastic change of leaving home for
boarding school. In choosing to accept the circumstances in which they found
themselves or by adopting a self-imposed system of emotional restraint they
had, as Mathew Thomson has observed of evacuees, 'to act, effectively, as their
own strict mothers and fathers'.[79] This mastery over difficult emotions was, in
part, what the sons of diplomatic families were going overseas to school to

learn. 'The English gentleman was ... a model of self-control ... "trained in self-repression, reticence and restraint"' wrote Marcus Collins (although notably describing his downfall). Public schools where, Collins states, the gentleman was educated, had a long history of inculcating boys with the specific emotional responses required from them.[80] When sociologists Lambert and Millham investigated boarding schools they remarked in detail on this phenomenon:

> In public schools and prep schools, for example, it is often the done thing towards the pupils not to display too much of one's feelings, never to be over-enthusiastic or behave like a spontaneous child. This norm arises partly from the school's ideal of educating a governing elite in which emotional reactions might be out of place, partly from early training in 'adult' attributes and partly from the hurt which the pupil might suffer by exposing his inner self to the gaze and criticism of his contemporaries.[81]

The perceived inevitability of boarding school, and the negative emotions that surrounded the ready-made decision to send children there, are made clear in a number of memoirs written by male diplomats whose families were young in the middle of the twentieth century. Bernard Burrows (1910–2002) recalled the moment his son was sent to prep school: 'We sent him there and he too duly got an Eton scholarship, but I am not sure that he enjoyed the preparatory school any more than I did.'[82] William Hayter (1906–1995) recalled that when his daughter Teresa was due to go to boarding school in 1949, from post in France, his Parisienne cook asked: 'Pauvre mademoiselle. Qu'est-ce qu'elle a fait?' because 'French children are not normally sent to a boarding school unless they are delinquent'.[83] It is implied that these families felt themselves to be powerless, especially in the case of Burrows. He did not enjoy his prep school, yet he sent his son there to undergo a similar experience. As Steiner put it, 'The pressure of the diplomatic environment is strong'.[84]

So integral was boarding school to Diplomatic Service culture that the cost (or at least a contribution towards it) was built into the job. Foreign Service Regulations from 1946 confirm that an annual Education Allowance of one hundred and fifty pounds was available to assist with boarding school costs for the first two children born into a family, although a survey of boarding school fees around this time suggests that this was far from adequate.[85] Also far from adequate were the allowances made for family travelling expenses according to the 1946 Regulations; unless a family travelled together, the 'officer will be responsible for the travelling expenses of his wife and family'.[86] One former diplomatic wife, interviewed in the 1980s, recalled that in the 1940s, 'My son aged ten flew to England from India and we did not see him again for a year – and then we had to pay his fare!'[87] It seems reasonable to assume, drawing on this remark, that during the post-war years, there was still an expectation that individual officers would have access to adequate private means.

Despite the apparent willingness of service families to accept family separation, the decision to send children away was not taken lightly. Olivia Tate⋆

remembered that her parents 'tried to avoid it for as long as possible' and the decision was recalled by Peter Boon as 'the dreaded business':[88]

> then came up to the dreaded business of what are we going to do? He's coming up to nine and so on, is it boarding school or not? And I suppose because my father had been to boarding school, [and] my mother, as an Indian civil service child, she had been to boarding school in India ... So they were both thoroughly familiar and knew all about boarding school.[89]

When he talked about his parents' boarding school decision Peter Boon immediately defended it with a series of justifications that became familiar from both children and parents during the course of this study. In many interviews, after the very painful decision to separate the family is discussed, a reason for this decision as the right one is immediately given. In Peter Boon's case (and in the case of many others) continuity was cited, by which Boon meant the continuity of education, something that was admittedly interrupted when children travelled with their parents: 'For the continuity, I think, for the continuity and ... I think it was very important really, yes.'[90] There remains a sense that at a fundamental level both children and parents were uncomfortable with the culture of separation and that they anticipated criticism from others by offering an immediate explanation.

Not only was the Foreign Office provision for school fees inadequate, a perennial problem existed of what to do with children during long holidays, especially in the summer. Securing holiday accommodation and care was far more of a problem during this period, when no travel allowances existed and air travel was not yet routine; children generally remained in Britain in the care of holiday homes or extended family. Holiday homes had a particularly bad reputation, crystallised by Rudyard Kipling in his story 'Baa Baa Black Sheep', which recounts the sorrowful experiences of an Anglo-Indian brother and sister, Punch and Judy, who remain as paying guests in the house of 'Uncle' Harry and 'Aunty' Rosa when their parents return to India.[91] Referred to by Buettner as 'The Kipling Paradigm', stories like 'Baa Baa Black Sheep' have provided a model for successive childhood memories of colonial life (memoirs and accounts refer to the story or make a direct comparison with it).[92] Peter Boon echoed this popular belief when he observed that 'many children of Empire were sent home to be educated and had to be farmed out during the holidays. Some of them ended up in sort of awful families and places' (although Peter had, however, no direct experience of either).

For the majority, extended family members in Britain remained the principal network of support. Deborah Cohen has noted the 'intimacy empire demanded of families', and while Cohen refers initially to the aid families in England could provide in terms of business transactions, their role soon came to encompass providing family members, especially children born overseas, with a home from home. 'Those who remained in the empire called upon relatives to raise their children ...' Cohen writes.[93] Both Peter Boon and Olivia Tate*, born in 1942 and 1946 respectively (and thus ineligible for the financial assistance provided to later families by the Plowden reforms), were cared for by grandparents during the long summer holidays. They enjoyed themselves but expressed doubts as to whether their relatives felt the same. Boon spent holidays with his grandmother:

> I think perhaps I was very lucky that the relatives were always there, including grandmother, she went back to Guernsey so there would be a little excitement there of flying over to Guernsey for the holidays ... I think my grandmother must have found it a bit difficult ... how to keep a ten, eleven, twelve year old occupied for, you know, four weeks![94]

Olivia Tate★ also spent the holidays with her grandmother who had a sure way of keeping her occupied:

> She used to take me to the office with her. She worked for an engineering company that made compressors ... but she managed to keep me busy. I was paid a salary and I used to deliver notes and deliver all sorts of things all around the offices ...[95]

Occasionally children travelled to see their parents, drawing on reserves of the expected independence and self-reliance that they were required to cultivate. Olivia Tate★ remembered the journey she took to see her parents in Poland in the late 1950s:

> going to Poland was interesting because most of the time we went by boat because Dad was not in Warsaw ... so one of the nuns would take me to Mark Brown's wharf and put me on this cargo ship and two days later I arrived in [Poland].[96]

If travel plans went wrong, children were expected to accommodate the change of schedule without anxiety: 'the first time in Karachi the plane ... one of the engines was faulty so we had an enforced twenty-four, thirty-six hour stay, and we had great fun being taken around Karachi and watching the crew play cricket'.[97] From time to time children's mobile lives offered them the opportunity to express themselves playfully and act boisterously. On one aircraft from Thailand, Peter Boon remembered a 'pillow fight and the senior steward came along and read the riot act ...'[98]

Another difficult holiday was at half term, a break which was until the early 1970s made up of only a day or an afternoon.

> In the fifties there weren't things like half term breaks ... and there weren't many occasions when parents came to the school ... Nevertheless there were opportunities to go out with parents for the day, and because the parents weren't in the country the opportunities to go out with the

family for the day was very limited ... One was sometimes invited by friends to go with them but it wasn't the same as having your own parents being with you and doing things with you.[99]

As children grew older they often found themselves in a position of responsibility towards their siblings, calling to mind Thomson's remark that evacuees had to take on a parental role.[100] Olivia Tate⋆ felt that she was 'generally sort of somewhere between big sister and mum' when she acted as guardian for her sister who was eleven years younger.[101] A number of commercial agencies existed to assist unaccompanied children through the initial stages of travel and gave themselves names associated with family, perhaps to suggest reliability or an intimate, personal quality of care, although the reality might have been very different. Among these were the Country Cousins Agency, based in West Sussex, although best known were the Universal Aunts, a company established in 1921, whose founder Gertrude Maclean had cared for young members of her own family while their parents were out of the country on Empire business. The Aunts offered a wide range of services and, in the aftermath of the First World War, attracted unmarried women and widows in need of respectable employment. Their promotional literature contained the suggestion that boarding school and its specific demands were a recognisable part of an average family's life and the tone of their advertisements implied that they shared upper middle class background with the people they set out to help. Many well-to-do women worked as 'Aunts' and throughout the period covered by this book the Foreign Office referred to their services as though they were an integral part of its community.[102]

Conclusion

This chapter began the task of describing what life was like for families who belonged to the British Diplomatic Service in the period directly following the Second World War and into the 1950s. The first section included a discussion of the way diplomats, in the years preceding the war, were still nostalgic for a time when the Service was 'a part of the King's household' and closely guarded the perceived social status of its members. High profile diplomatic families like that of Duff and Diana Cooper were contrasted with the intellectual and close-knit Dixons, who nonetheless quickly founded themselves assimilated into the Foreign Office's elevated social circles and by extension into British high society. The second section viewed the diplomatic family through the lens of post-war reconstruction in Britain and its home-centred social reforms, crucially noting that it was at this point that the experiences of Diplomatic Service children began to diverge from those reasonably expected by their peers in the UK. Despite the fact that they experienced far greater material comfort overseas, it was seen that this comfort was a veneer that obscured a number of far more challenging demands, including a far greater expectation of responsibility from children and young people and a lack of privacy, sometimes embodied by

Soviet spies posing as servants. This section also introduced diplomatic children's confused notions of 'home' both as a material place and a 'homeland' and placed this confusion alongside the experiences of similar groups like army children and the children of colonial administrators. It could be seen immediately, then, that Foreign Office practice diverged from the post-war ideal of the family; and that its children were largely unable to enjoy the post-war pleasures of home and family 'idealised as all that children needed' as Thomson has said, owing to the entrenched culture of family separation which existed in the Diplomatic Service for reasons of tradition and loyalty and which was, from 1946, subsidised by the Service itself. Separation was one of the most enduring, complex and affecting characteristics of the lives of British diplomatic children: one which, remarkably, was still common practice at end of the period covered by this study. This raises questions about the position of diplomats' families with regard to Britain's updated social customs and asks how far they could be said to authentically represent the UK when their own lives were becoming so different.

Notes

1 Jennifer Mori, *The Culture of Diplomacy* (Manchester: Manchester University Press, 2010), 17–18.

2 Valerie Cromwell, 'A World Apart: Gentleman Amateurs To Professional Generalists', in *Diplomacy and World Power Studies in British Foreign Policy 1890–1950*, ed. Michael Dockrill and Brian McKercher (Cambridge: Cambridge University Press, 1996), 2.

3 Marcus Cheke, *Guidance on foreign usages and ceremony, and other matters, for a member of His Majesty's Foreign Service on his first appointment to a Post Abroad* (London: Foreign and Commonwealth Office, 1949). Anne Coles, 'Making Multiple Migrations: The Life of British Diplomatic Families Overseas', in *Gender and Family among Transnational Professionals*, ed. Anne Coles and Anne-Meike Fechter (London: Routledge, 2008), 28–129, devotes a section to the 'Mission as family' and speculates that this 'patriarchal and hierarchical' notion has endured and been useful into the twenty-first century in small posts.

4 Masha Williams, 'Foreign Service Wives Association: How It Started', *Diplomatic Service Wives Association Newsletter*, Autumn 1985 Anniversary Edition, 52–53.

5 Coles, 'Making Multiple Migrations', 139.

6 David Cannadine, *The Decline and Fall of the British Aristocracy* (New Haven and London: Yale University Press: 1990), 280.

7 Desmond Platt, *The Cinderella Service: British Consuls since 1815* (London: Longman, 1971), Appendix 240–242. When the letter was written Knatchbull-Hugessen was British Ambassador in Istanbul.

8 Proposals for the reform of the Foreign Service Presented by the Secretary of State for Foreign Affairs to Parliament by Command of his Majesty, 1943. Cmnd. 6420, Introduction, 2, 214.

9 Proposals for the Reform of the Foreign Service, White Paper of January 1943, *Cmd 6420* (hereafter, 1943 White Paper). Introduction, 7.

10 Cannadine, *Decline and Fall of the British Aristocracy*, 280.

11 Cannadine, *Decline and Fall of the British Aristocracy*, 232; Andrew Roberts, *Holy Fox: The Life of Lord Halifax* (London: Head of Zeus, 2014), 3; Rohan Butler 'Lord Halifax', *Diplomatic Service Wives Association Newsletter*, Autumn 1985 Anniversary Edition, 62.

12 Roberts, *Holy Fox*, 33, 12.
13 David Kelly, *The Ruling Few* (London: Hollis and Carter, 1952), 118.
14 Marie-Noele Kelly, *Dawn to Dusk* (London: Hutchinson, 1960), 117.
15 Ibid, 155.
16 Ibid, 32.
17 Ibid.
18 Ibid.
19 Knatchbull-Hugessen letter in Platt, *The Cinderella Service*, 241.
20 According to the OED the word 'parenting' originated in 1918 in the US.
21 Kelly, *Dawn to Dusk*, 155.
22 Cannadine, *Decline and Fall of British Aristocracy*, 284.
23 John Julius Cooper to his mother, Lady Diana Cooper, July 1940 in John Julius Norwich (ed.), *Darling Monster: The Letters of Lady Diana Cooper to Her Son John Julius Norwich 1939–1952* (London: Vintage, 2013), 30.
24 Olivia Tate★. Interview by Author. 10 March 2014.
25 Piers Dixon, *Double Diploma. The Life of Sir Pierson Dixon Don and Diplomat* (London: Hutchinson and Co., 1968), 2
26 Ibid, 10.
27 James Ellison, 'Pierson Dixon 1960–1965', in *The Paris Embassy, British Ambassadors and Anglo-French Relations*, ed. Rogelia Pastor-Castro and John W. Young (Basingstoke: Palgrave MacMillan, 2013), 93.
28 *The Double J, May 1941*, Children's Manuscript Magazine, Dixon Family Archive.
29 Ibid.
30 Dixon, *Double Diploma*, 246.
31 Cromwell, 'A World Apart', 9.
32 Both Helen McCarthy, *Women of the World: The Rise of the Female Diplomat* (London: Bloomsbury, 2014), 91–92, and Cromwell, 'A World Apart', 3–4, discuss the effects of the Commission's findings on Foreign Office culture.
33 *1943 White Paper*. Introduction: 4, 215; Introduction: 2, 214.
34 *1943 White Paper,* Introduction: 4, 214.
35 Jean Schofield Heywood, *Children in Care. The Development of the Service for the Deprived Child* (London: Routledge, 2013), 133.
36 Mathew Thomson, *Lost Freedom: The Landscape of the Child and the Postwar Settlement* (Oxford: Oxford University Press, 2013), 10.
37 Michal Shapira, *The War Inside: Psychoanalysis, Total War and the Making of the Democratic Self in Post-War Britain* (New York: Cambridge University Press, 2013), 202.
38 See Hugh Cunningham, *The Invention of Childhood* (London: BBC Books, 2006): on children and the welfare state (178); on the tensions between 'behaviourists' and advocates of a more relaxed form of discipline (198–201); on 'investment' in children, emotional and otherwise (215).
39 Shapira, *The War Inside*, 203.
40 Ibid., 129.
41 Thomson, *Lost Freedom*, 225.
42 Claire Langhamer, 'Meanings of Home in Postwar Britain', *Journal of Contemporary History* 40:2 (2005): 341–362, 345.
43 Ibid., 202.
44 'The century of the child' is often used to refer to developments in thinking about the situation of children during the twentieth century and was taken from title of a book by Swedish educationalist Ellen Key which was translated into English as *The Century of the Child* and published in 1909.
45 Peter Boon. Interview by Author. 11 April 2014.
46 Kelly, *Dawn to Dusk*, 155–156.
47 Questionnaire. Denise Holt. 26 July 2015

48   Peter Boon. Interview by Author. 11 April 2014.
49   Olivia Tate★. Interview by Author. 10 March 2014.
50   Penny Summerfield, 'Women in Britain since 1945', in *Understanding Post-war British Society*, ed. James Obelkevich and Peter Catterall (London: Routledge, 1994), 61.
51   Kelly, *Dawn to Dusk*, 116; 234.
52   Peter Boon. Interview by Author. 11 April 2014.
53   Kelly, *Dawn to Dusk*, 209.
54   Olivia Tate★. Interview by Author. 10 March 2014.
55   Katie Hickman, *Daughters of Britannia: The Life and Times of Diplomatic Wives* (London: Flamingo, 1999), 73.
56   Helen McCarthy, *Women of the World: The Rise of the Female Diplomat* (London: Bloomsbury, 2014), 256.
57   Coles, 'Making Multiple Migrations', 129.
58   Peter Boon. Interview by Author. 11 April 2014.
59   Mori, *Culture of Diplomacy*, 17.
60   See, for example, McCarthy, *Women of the World*, 31–32, for the development of this phenomenon.
61   Harrison, *Seeking a Role*, 190, 213.
62   Elizabeth Buettner, *Empire Families* (Oxford: Oxford University Press, 2004), 189.
63   Stuart Hall, 'Minimal Selves', in *The Real Me: Post Modernism and the Question of Identity*, ed. Lisa Appignanesi, ICA documents 6 (London: The Institute of Contemporary Arts, 1987), 44–46. Coles, 'Making Multiple Migrations', 128–129.
64   Denise Holt. Questionnaire. 26 July 2015.
65   Mike Ingalls★. Questionnaire. 6 October 2014
66   Elizabeth Buettner, 'Parent-Child Separations and Colonial Careers: The Talbot Family Correspondence in the 1880s and 1890s', in *Childhood in Question: Children, Parents and the State*, ed. Anthony Fletcher and Stephen Hussey (Manchester: Manchester University Press, 1999), 125.
67   See Harrison, *Seeking a Role*, 265.
68   Denise Holt. Questionnaire. 26 July 2015.
69   Donald L. Sampson and Howard P. Smith, 'A Scale to Measure World-Minded Attitudes', *The Journal of Social Psychology* 45:1 (1957): 99–106.
70   Peter Boon. Interview by Author. 11 April 2014.
71   Olivia Tate★. Interview by Author. 10 March 2014.
72   David Pomfret, *Youth and Empire: Trans Colonial Childhood in British and French Asia* (Stanford: Stanford University Press, 2015).
73   Mori notes that by 1780 'a suitable wife was starting to be seen as a necessary partner' (*Culture of Diplomacy*, 70) and 35 years later in 1815 'marriage had become pre-requisite for senior posts' (71). Mary A. Procida, *Married to the Empire: Gender Politics and Imperialism in India, 1883–1947* (Manchester: Manchester University Press, 2002), 13.
74   Buettner, *Empire Families*, 29; Pomfret, *Youth and Empire*, 24.
75   Richard Crossman, *Palestine Mission* (New York: Harper, 1947), 131.
76   Buettner, *Empire Families*, 72–180. This issue is covered extensively.
77   Ibid., 107.
78   Ibid., 110–145.
79   Thomson, *Lost Freedom,* 65–66.
80   Marcus Collins, 'The Fall of the English Gentleman', *Historical Research* 75:187 (2005): 90–111.
81   Royston Lambert with Spencer Millham, *The Hothouse Society. An Exploration of Boarding School through the Boys' and Girls' Own Writing* (Harmondsworth: Penguin, 1974), 224–225.

82 Bernard Burrows, *Diplomat in a Changing World* (Spennymoor County Durham: The Memoir Club, 2001), 6.
83 William Hayter, *A Double Life* (London: Hamilton, 1974), 95.
84 It could be argued that the end justified the means. Burrows Jnr 'duly got his scholarship to Eton'. Zara Steiner, 'The Foreign and Commonwealth Office: Resistance and Adaptation to Changing Times', *Contemporary British History* 18:3 (2004): 13–30, 23.
85 TNA, FO 366/1644. Buettner, *Empire Families*, 166, discusses the cost of boarding school fees, quoting the fees charged at a 'mid-range' school, Cheltenham College, which were between £84 and £103 for boarders in 1912. Eton and Harrow – schools at the top of the range, the former often favoured by Foreign Office – charged £166 or £153 respectively in the same year. The annual fee for a boarder at Dulwich College in South London in 1949 was £104. It seems reasonable to suppose that fees had risen considerably thirty-four years later and that the Foreign Office allowance did not therefore cover the annual boarding school fee for a 'mid-range' school in its entirety.
86 TNA, FO 366/1644.
87 Anonymous questionnaire, 1980s. Smedley Box A.
88 Peter Boon. Interview by Author. 11 April 2014.
89 Ibid.
90 Ibid.
91 Rudyard Kipling, 'Baa Baa Black Sheep', in *The Man Who Would be King and Other Stories* (Oxford: Oxford World's Classics, 1999), 170–197.
92 Buettner, *Empire Families*, 121–129.
93 Deborah Cohen, *Family Secrets: Living with Shame from the Victorians to the Present Day* (London: Penguin, 2013), 7.
94 Peter Boon. Interview by Author. 11 April 2014.
95 Olivia Tate★. Interview by Author. 10 March 2014.
96 Olivia Tate★. Personal Interview. 10 March 2014. Tate's★ father was Consul in regional Poland. When the author attended a conference on unaccompanied children in June 2014 she discovered that a Jewish refugee child (the subject of another paper) had made Tate's★ exact journey in reverse shortly before the Second World War.
97 Peter Boon. Interview by Author. 11 April 2014.
98 Ibid.
99 Ibid.
100 See footnote 29 above.
101 Olivia Tate★. Personal Interview. 10 March 2014.
102 See Kate Herbert-Hunting, *The Universal Aunts* (London: Constable, 1986).

Bibliography

Buettner, E. *Empire Families* (Oxford: Oxford University Press, 2004).

Buettner, E. 'Parent-Child Separations and Colonial Careers: The Talbot Family Correspondence in the 1880s and 1890s', in *Childhood in Question. Children Parents and the State*, ed. A. Fletcher and S. Hussey (Manchester: Manchester University Press, 1999).

Burrows, B. *Diplomat in a Changing World* (Spennymoor Country Durham: Memoir Club, 2001).

Cannadine, D. *The Decline and Fall of the British Aristocracy* (New Haven and London: Yale University Press, 1990).

Cheke, M. *Guidance on Foreign Usages and Ceremony, and Other Matters, for a Member of His Majesty's Foreign Service on His First Appointment to a Post Abroad* (London: Foreign and Commonwealth Office, 1949).

Cohen, D. *Family Secrets: Living with Shame from the Victorians to the Present Day* (London: Penguin, 2013).

Coles, A. 'Making Multiple Migrations: The Life of British Diplomatic Families Overseas', in *Gender and Family Among Transnational Professionals*, ed. A. Coles and A. Fechter (London: Routledge, 2008).

Collins, M. 'The Fall of the English Gentleman: The National Character in Decline, c.1918–1970', *Historical Research* 75:187 (2002).

Cromwell, V. 'A World Apart: Gentleman Amateurs to Professional Generalists', in *Diplomacy and World Power Studies in British Foreign Policy 1890–1950*, ed. M. Dockrill and B. McKercher (Cambridge: Cambridge University Press, 1996).

Crossman, R. *Palestine Mission* (New York: Harper, 1947).

Cunningham, H. *The Invention of Childhood* (London: BBC Books, 2006).

Diplomatic Service Wives Association (DSWA) Newsletter, Autumn 1985 Anniversary Edition.

Dixon, P. *Double Diploma. The Life of Sir Pierson Dixon* (London: Hutchinson, 1968).

The Double J, Children's Manuscript Magazine, 1942–1947. Dixon Family Archive.

Ellison, J. 'Pierson Dixon 1960–1965', in *The Paris Embassy, British Ambassadors and Anglo-French Relations*, ed. R. Pastor-Castro and J.W. Young (Basingstoke: Palgrave MacMillan, 2013).

FO File 366/1644 The National Archives, Kew, London.

Hall, S. 'Minimal Selves', in *The Real Me: Post Modernism and the Question of Identity*, ed. L. Appignanesi, ICA documents 6 (London: The Institute of Contemporary Arts, 1987).

Harrison, B. *Seeking a Role: The United Kingdom 1951–1970* (Oxford: Clarendon, 2009).

Hayter, W. *A Double Life* (London: Hamish Hamilton, 1974).

Heywood, J.S. *Children in Care The Development of the Service for the Deprived Child* (London: Routledge, 2013).

Herbert Hunting, K. *The Universal Aunts* (London: Constable and Partners, 1986).

Hickman, K. *Daughters of Britannia: The Lives and Times of Diplomatic Wives* (London: Flamingo, 1999).

Kelly, D. *The Ruling Few* (London: Hollis and Carter, 1952).

Kelly, M. *Dawn to Dusk* (London: Hutchinson, 1960).

Kipling, R. 'Baa-Baa Black Sheep', in *The Man Who Would be King and Other Stories* (Oxford: Oxford World's Classics, 1999).

Lambert, R. with Millham, S. *The Hothouse Society. An Exploration of Boarding School Life through the Boys' and Girls' Own Writings* (London: Weidenfeld and Nicolson, 1968).

Langhamer, C. 'Meanings of Home in Postwar Britain', *Journal of Contemporary History* 40:2 (2005).

McCarthy, H. *Women of the World: The Rise of the Female Diplomat* (London: Bloomsbury, 2014).

Mori, J. *The Culture of Diplomacy* (Manchester: Manchester University Press, 2010).

Norwich, J.J. ed. *Darling Monster. The Letters of Lady Diana Cooper to Her Son John Julius Norwich 1939–1952* (London: Vintage, 2013).

Platt, D. *The Cinderella Service: British Consuls since 1815* (London: Longman, 1971).

Pomfret, D. *Youth and Empire: Trans Colonial Childhood in British and French Asia* (Stanford: Stanford University Press, 2015).

Procida, M.A. *Married to the Empire: Gender Politics and Imperialism in India 1883–1947* (Manchester: Manchester University Press, 2002).

Proposals for the reform of the Foreign Service Presented by the Secretary of State for Foreign Affairs to Parliament by Command of his Majesty, 1943. *Cmnd.* 6420.

Roberts, A. *Holy Fox: The Life of Lord Halifax* (London: Head of Zeus, 2014).

Sampson, D. and Smith, P. 'A Scale to Measure World-Minded Attitudes', *The Journal of Social Psychology* 45:1 (1957).

Shapira, M. *The War Inside: Psychoanalysis, Total War and the Making of the Democratic Self in Post-War Britain* (New York: Cambridge University Press, 2013).

Steiner, Z. 'The Foreign and Commonwealth Office: Resistance and Adaptation to Changing Times', *Contemporary British History* 18:3 (2004): 13–30.

Summerfield, P. 'Women in Britain since 1945', *in Understanding Post-war British Society*, ed. J. Obelkevich and P. Catterall (London: Routledge, 1994).

Thomson, M. *Lost Freedom: The Landscape of the Child and the Postwar Settlement* (Oxford: Oxford University Press, 2013).

2   1958–1971

In February 1964 the British pop group The Beatles arrived in America. Such was their popularity on that side of the Atlantic that they were greeted at Kennedy Airport by ten thousand screaming fans and a record seventy-three million viewers watched their appearances on *The Ed Sullivan Show.* [1] During their American tour, The Beatles attended a diplomatic reception at the British Embassy in Washington. According to their biography: 'Officious junior officials started pushing them around, insisting they spoke to people and gave autographs. "Sign this," one said to John [Lennon], who refused. "You'll sign this and like it."'[2] This account of The Beatles' experience at a diplomatic party says much about the rapid changes overtaking British society in the 1960s and perhaps just as much about the way the Diplomatic Service positioned itself in relation to those changes. The Beatles' popularity made them perfect 'ambassadors' for the UK: their reception proved that they were a tremendously successful export. However, their irreverence and candour meant that they represented a new expression of 'Britishness', one that did not coincide with the 'intangible qualities' beloved by diplomats like Hugh Knatchbull-Hugessen and formerly demanded by the Service of its representatives. The arrogance of the 'junior official', who told John Lennon that he would sign an autograph 'and like it', was rendered somewhat ridiculous by The Beatles' self-possession. Popular artists like The Beatles symbolised the self-confident, less deferential spirit of the new decade: this autocratic expectation of obedience and conformity was out of date. Another member of the band, George Harrison, later reflected on that evening and others like it: 'They were always full of snobby people who really loathe our type, but want to see us because we're rich and famous. It's all hypocrisy.'[3]

This unsatisfactory encounter between a vibrant new British cultural export and Britain's existing 'representational services overseas' provided evidence of how the Diplomatic Service did not always move with the times, especially when those times seemed to be diametrically opposed to Foreign Office traditions. The 1960s celebrated the individual while established institutions lost respect; trust and good manners were no longer prized, along with 'correct' forms of dress and categories of social class. However, during this period, the lives of Foreign Office personnel and their families were transformed whether

DOI: 10.4324/9780429273568-2

they resisted or not by demands that grew out of significant worldwide changes. The rapid dismantling of Britain's colonies in Africa and South Asia meant that diplomats were forced to adapt to a number of new – and no longer deferential – governments and posts; while the diminishing status of the UK as a world power led to internal restructuring both in the Foreign Office and the wider civil service. During this era Foreign Office families – and especially diplomatic wives – felt able to challenge the Foreign Office status quo for the first time, perhaps inspired by these changes. This resulted in momentous change for the families of diplomats and gave them the confidence to organise their demands and take them further.

This chapter examines the effects of the changing political and social landscape of the 1960s on Foreign Office families. It will seek to determine whether the recommendations of the 1943 White Paper which proposed the recruitment of a more diverse diplomatic workforce had been successfully put into effect. By the 1960s a generation of recruits that had benefitted from the Attlee government's educational reforms and from subsequent expansions in higher education after the 1963 Robbins Report began to create diplomatic families of their own. Less willing to accept existing conditions, the attitudes of this new generation can be explored through a close examination of government and Foreign and Commonwealth Office (FCO) internal policy towards families. The first section explores the evidence concerning diplomatic family life which was submitted by members of the Foreign Service Association to the 1964 Plowden Report on Representational Services Overseas. They were aided in this by members of the Foreign Service Wives Association that had been formed in 1961 to assist diplomats' wives with the singular problems that confronted them. Both groups were triumphant after the report recommended an extension of allowances related to children's travel, meaning that families could spend more time together. The second section looks at youth culture and asks how far evidence suggests that diplomats' children entered into the events and spirit of the 1960s through its music, politics and 'counter-culture'. It will also trace the beginnings of the 'third culture', so-called by anthropologists and rapidly endorsed as an alternative identity for children growing up internationally and 'on the move' in the post-colonial world. The final section of this chapter returns to an assessment of family circumstances within the Foreign Office and the battle to build on family-friendly allowances for Diplomatic Service children initially set in motion by the findings of the Plowden Report. It draws on primary evidence, kept in the British National Archives, of a campaign by diplomats to secure a third 'concessionary' journey to enable their children to travel to and from post in the school holidays.

The Plowden Report

Sally James, young wife of a diplomat, wrote in March 1964:

> The recent Plowden report, which is a review headed by Lord Plowden, on the workings of the Foreign Office, is very good. At last someone

seems to understand the problems of having to move from post to post while trying to give children a settled education ...[4]

This section will explore the impact of the 1964 *Report of the Committee into Representational Services Overseas* which was seen as so beneficial to diplomatic families that it was referred to simply as 'Plowden'. Plowden enabled diplomats who had joined the Foreign Office after the 1943 reforms to demand larger and better targeted allowances and to voice their dissatisfaction with existing practices regarding family separation. Additionally, the Plowden Report was the first investigation into Foreign Office practices to which diplomatic wives were able to contribute directly. This they did through the newly formed Foreign Service Wives Association which, since its formation in 1960, had been working to improve the lives of diplomatic wives at home and abroad. There can be little doubt that the Plowden Report was a landmark in the fortunes of the Foreign Office family; its status as a milestone is such that the researcher will sometimes find sources that refer simply to Pre- and Post-Plowden.[5]

The Plowden Report was commissioned by Prime Minister Harold Macmillan in 1962, as Britain's global role was undergoing rapid and major change. Throughout the 1960s, the UK faced the dismembering of its African colonies; financial obligations to the US following the Second World War had led to British involvement in Cold War politics; while relations with Europe were strained, a second British application to join the European Economic Community (EEC) was vetoed by France in 1963. On the national stage, traditional social hierarchies were being broken down. In 1958, for example, the last cohort of debutantes were presented at court. These cultural shifts were also being recognised at the theatre in 'kitchen sink dramas' such as *Look Back In Anger* (1956) and in the novels of Alan Sillitoe and Stan Barstow which celebrated working class lives. To a great extent, however, the Foreign Office, always 'a world apart', managed to resist these changes and maintained allegiance to an archaic, courtly system. Cromwell notes 'the considerable assumption that established traditions and practices would continue ...'[6] The wider implication of this institutional mindset was that new staff would have to adjust their behaviour to suit Foreign Office practice.

By the time the Plowden Committee began to gather evidence, the Diplomatic Service entrants who had benefitted two decades earlier from the 1943 White Paper's scheme of 'special entry' based on distinguished military service and from the subsequent 'meritocratic and democratic' education policy of the post-war Labour government were establishing themselves as diplomats, marrying and starting families.[7] Harrison described 'the hard-working grammar school boy who did well at Oxford or Cambridge, took the civil service examination, thereafter progressing steadily into the elite'.[8] However, just as it became necessary for Marcus Cheke's book of etiquette to be updated to enlighten post-war recruits, the system and value of diplomatic allowances also appeared to need restructuring.

It was the intention of the Plowden Committee:

> To review the purpose, structure and operation of the services responsible for representing the interests of the United Kingdom government overseas and to make recommendations having regard to changes in political, social and economic circumstances in this country and overseas.[9]

Its scope encompassed the Foreign Service, the Board of Trade and the Commonwealth Relations Office and its lasting legacy in Whitehall was the recommended amalgamation of all three: to be known as HM Diplomatic Service. The initial meeting of the Sub Committee on Conditions of Service in November 1962 identified a number of topics for investigation; among them was 'Children'. The Sub Committee recognised that despite the fundamental principle of the 1943 White Paper, members of staff without private means were still experiencing financial hardship. During this early discussion it was noted that 'Allowances for children's education are still badly out of line with what is needed'.[10] The need to re-examine the number and cost of concessionary journeys was also mentioned, although the Treasury's representative, Mr Hunt, constantly urged caution, conscious that the distribution of allowances was 'an area of great political sensitivity'.[11] Presumably the Treasury was concerned about envious comparisons being made by members of the Home Civil Service and staff of other government departments whose circumstances did not qualify them for the same allowances as their diplomatic colleagues. There was also the need to manage public opinion.

One group that was more than happy to learn of the Plowden Committee's scrutiny was the Foreign Service Wives Association (FSWA). This body was the outcome of diplomatic wives' growing impatience with the dearth of pastoral care and other provision for 'camp followers' provided by the Service. It was conceived by a small and dedicated group who described its aims thus: 'Our first concern is to provide a centre to which our wives may bring their problems.'[12] A founder member, Masha Williams, felt 'that the strain on us all was intolerable' and the introduction to the wives' written evidence to the Plowden Committee stated that:

> it had been felt for some time that too little had been done in the past to help the wives and families of Foreign Service Officers with the particular problems that confront them due to the nomadic lives they lead.[13]

The founder members had felt initially, as they compiled a memorandum of problems and suggested solutions, 'like conspirators' and worried that they might be 'labelled bolshie'.[14] The high profile case of Diana Bromley, however, the wife of a diplomat who killed her two young sons and attempted to kill herself after a suffering a breakdown, that made British newspaper headlines in December 1958, helped to facilitate a more sympathetic attitude towards wives (as well as involving the Foreign Office in negative newspaper coverage, something often guaranteed to effect change) and no doubt contributed to the creation of the Association a year later. In an account of their first AGM in July 1961, President Lady Rundall mentioned a circular signed by John Henniker-Major, then Head of Personnel, which 'spoke of a need for special study of the problems which confront Foreign Service wives, and of the increasing need for thinking in terms of family welfare'.[15] This last quote reinforces what Walsh has termed 'the persistent gendering of expatriate lives' in the way that it automatically links wives' problems with those of family.[16] However, the Plowden Committee also received evidence from the Foreign Service Association (FSA), a trade union that represented Diplomatic Service Officers in the Foreign Office 'fast stream', Branch A, which provided a striking demonstration that children were as much a source of concern for husbands and fathers as for wives and mothers.[17]

A more detailed look into the FSWA's evidence to the Plowden Committee gives a clearer picture of the strains on Foreign Service family life that were particularly felt by the wives by the early 1960s. The predominate anxiety was the enforced periods of separation from children, but the financial hardship resulting from inadequate boarding school and travel allowances were regarded as adding insult to injury, especially among officers from less privileged social backgrounds, who lacked a private income. The 'post-1943' group may also have struggled with accepted family practices within the Foreign Office as they did not belong to and share the tradition of the upper class and colonial families, discussed in the final part of Chapter One, who were resigned to long periods of family separation. Indeed, the section of FSWA evidence to the Plowden Committee that dealt with the family contained the heading 'The Problem of Children', suggesting that the existence of children created for married couples a further area of stress in an already difficult situation. Certainly, the level of devotion demanded by the Foreign Service could be said to have created a tension which saw a mother forced to place her duty to her husband and his chosen career above her children. A talk entitled 'Serving Abroad' given to diplomatic wives in November 1960 by Lady Kirkpatrick, wife of the Permanent Under Secretary, echoing this sentiment, included this advice: 'I have chosen the title *Serving Abroad* because *service* is the key note: and if we realise that the Service is more important than we are, we shall do our work abroad properly.'[18]

When it came to give evidence to the Plowden Committee, the FSWA asserted that 'All responsible parents are deeply disturbed by the enforced separation from their children'.[19] They understood, they said, the practical reasons why boarding school was often more suitable for older Diplomatic Service children:

> firstly to keep up with the demands of an increasingly higher educational standard at home, secondly, to provide some continuity and stability, thirdly, as an insurance in case they are sent to a post where there are no educational facilities.[20]

Yet the FSWA rejected the impositions that were understood, in Lady Kirk-patrick's address, to be an integral part of diplomatic life. They knew that they, as wives, would not be entirely exempt but were not prepared to accept them on behalf of their children.

> It could be argued that, as the parents chose the Foreign Service life, they must make the sacrifice involved ... but in this case it is the children who, through no fault of their own, have to pay the penalty.[21]

The views expressed by the FSWA members reflected neither the English middle class tradition of sending children to boarding school, nor the colonial family's sense of 'parental sacrifice', which had, admitted the Commonwealth Relations Office representative to the Plowden Committee, 'permeated the whole system'.[22] It would also be an over-simplification to assume that this point of view belonged to an entirely 'new' and 'younger' diplomatic genera-tion. With the Foreign Office's enthusiasm for hierarchy in mind, it can rea-sonably be assumed that many FSWA members, especially those occupying higher positions, would be older wives of more established diplomats who had had greater experience of separation and were likely to hold less liberal attitudes to family life. The relaxation of attitudes towards parental discipline and growing disapproval of family separation against the background of childcare advice from 'experts' like John Bowlby and Donald Winnicott, discussed in the preceding chapter, may also have informed the wives' position, or created a more sympathetic atmosphere for their expression.[23] In the light of this 'enforced' separation, the FSWA drew the Committee's attention to the importance of school holidays for Foreign Service children, and the following request was made:

> The only satisfactory solution is for the children to rejoin their parents for each holiday. Yet without financial help this is virtually impossible for all those whose fathers are serving some distance from the United Kingdom. For Her Majesty's Government to pay the air fares of children rejoining their parents three times a year instead of once would be a most valuable reform.[24]

The Foreign Service Association testimony, submitted in the autumn of 1962, was far more direct in its criticisms and demands. An outspoken document, it listed a number of grievances which had ostensibly been taken from letters (anonymised for the report) sent to the Association over a period of time. Possibly, FSA mem-bers felt (in common with the wives when they formed their Association) that a low level of subterfuge was needed in case their negative comments led to them being perceived as 'bolshy' and became detrimental to career success. Many of the comments recorded in the FSA document pointed towards financial hardship. Its section on children began by noting that 'the problems of raising and educating a family under the conditions of life in the Service is the one which worries more members than any other' and continued 'the problem threatens to become intolerable when this is coupled with inadequate financial provision'.[25] Zara Steiner's observation that 'Plowden in 1963 ... still found differences and distinctions between types of post and the kinds of people who filled them...' is striking in the context of the FSA's frequent referral to colleagues with private means.[26] It points to an abiding tension between diplomats who were privately wealthy and those from less affluent backgrounds. Some of these comments are heavy with sarcasm, like the following footnote about the word 'concessionary' that described children's 'concessionary' holiday journeys:

> The word 'concessionary' is indicative. The impartial observer might think it to imply that Foreign Service Officers had no business, unless they had private means, to expect to see their children of school age at any time in the course of a tour of duty abroad, and that the allowance was provided with some reluctance.[27]

Elsewhere FSA members expressed the view that the insufficiency of travel, boarding school and accommodation allowances, leading to an inability to fund a family on a diplomatic salary and allowances, could prove a significant barrier to career mobility and ultimately force a change of career. The financial concerns faced by the diplomats who submitted evidence to the Plowden Committee via the FSA may also have added to the emotional burden experienced by diplomatic parents. The perceived inability of male diplomats to provide for their own families was potentially very wounding in terms of their self-image as parents.[28] As the work of Laura King on fatherhood has shown, a growing emphasis on fathers taking pleasure and pride in their role as family provider began to develop from the 1930s and was strengthened by the emotional intensity of family separations during the war years.[29] King's claim that fathers' emotional involvement with their children also increased during this period would have presented Foreign Office fathers with difficulties in the face of enforced separation.[30] As the parent who was always the official employee of the FCO in this period, fathers did not have as much opportunity to provide emotional support to their children, something that was available to their wives, who were, for instance, able accompany children to boarding school to help them settle in. The feeling of distance that diplomatic children often felt between themselves and their fathers was enhanced by these regulations and the paucity of financial assistance.

The FSA paper exposes a set of highly emotional reactions towards the inevitable separations diplomatic family life entailed and the notion of parental sacrifice. The all-encompassing character of the Foreign Office is criticised: 'The Senior Branch of the Foreign Service requires that the officer and his wife should virtually devote their lives to the Service ... The arrangements for their children's upbringing must be such as the service permits.'[31] An arresting footnote to the FSA paper stated:

the number of 'problem children' of Foreign Service officers who would not have been problem children if the father's vocation had kept him permanently in the United Kingdom is ... not susceptible of ... arithmetical assessment but it is ... nonetheless disturbing.[32]

The use of the label 'problem children' demonstrates both the growing influence of psychology in the understanding of family welfare and the awareness that Diplomatic Service life was not always good for children.[33] This sense that diplomats' children could be adversely affected by their unsettled lives resurfaces in other writing about the Foreign Office. Hickman, for instance, reported 'Counsellors at the Family Welfare Department at the Foreign Office had seen increasing numbers of diplomatic children coming to them for help'. Later accounts of life at the Foreign Office also made this connection between the diplomatic lifestyle and mental health problems in children, and whilst the evidence for this is inconclusive, Diplomatic Service children did experience periods of unhappiness and were forced to cope with greater demands than their peers outside the Diplomatic Service.[34]

Considering the evidence collected, the Sub Committee was no doubt well aware of the effect of family matters – especially boarding school allowances and concessionary journeys – on Foreign Office staff. After discussing the paper submitted by the FSA, the Sub Committee 'considered that these matters were crucial from the point of view of the efficiency and morale of overseas services'.[35] Morale, that of both the adults and children, was mentioned regularly throughout the minutes. The disheartening effects separation was felt to have on children was of great concern to the FCO officer and his wife and, as the Sub Committee saw it, hindered them in carrying out their duties. During the first Sub Committee meeting of 1963, HM Treasury's representative 'said that the Treasury was sympathetic and hoped that they may be able to do something on local school fees'.[36] However, the fear of repercussions at both a professional (that is, among other civil servants and related occupations) and national level meant that they were forced to perform a delicate balancing act. A Treasury note from April 1963 mentions that 'any change to the Boarding School Allowance would almost inevitably have to be followed by a similar increase in the allowances of the Home Civil Service abroad and the Armed Forces'.[37]

Although it was not made explicit, the Plowden Committee must have also been concerned for public opinion. The wives prefaced their evidence with 'the impression which still manages to persist [is] that life in the Foreign Service is one of glamorous ease'.[38] This section has attempted to demonstrate the considerable financial hardship suffered by the post-1943 recruits yet some aspirational newcomers were keen to give the impression that they had joined and were representing an elite and leisured class. This was perhaps due to the process of 'social acculturation' involved in entry to the Foreign Office that was identified by both Cromwell and Steiner. Despite protestations to the contrary, members of the Diplomatic Service could not always help talking with a flourish about their social connections, grand accommodation or domestic help and this must have influenced the way they were seen by outsiders. It was vital, then, that the Committee was not seen to be providing benefits to an institution already perceived to be privileged.

Ultimately the Treasury took the view that an overhaul of concessionary journeys should be the 'main feature of the settlement of problems relating to children'.[39] The considerations discussed above were carefully outlined in the final Plowden Report, published in February 1964. It recommended a high level of practical help for Foreign Service families. The summary of principal conclusions set out in the report's Conclusion observed succinctly: 'A new system of boarding school allowances is required. The levels of these allowances should be raised substantially ...'[40] Difficulties posed by the cost of long distance travel were also reviewed and a second annual 'concessionary' journey for children at school was granted.[41] These two recommendations, read the Report, 'are designed as an integral whole ... We believe that their acceptance and implementation should be regarded as matters of importance and urgency.'[42] The Committee struck a balance between a display of sympathy with diplomatic parents and the need to defend its decisions to critics within the civil service and the wider public domain. 'The Foreign Service career', it explained, 'is governed by a number of factors which collectively and cumulatively warrant special treatment.'[43] The question of how far the Foreign Office was concerned with retaining the staff it recruited is demonstrated in the Plowden Report. It was to show the government's dedication towards creating a more equitable Foreign Service, of the kind outlined in the 1943 White Paper: 'there will be a danger that men without private means will be deterred for financial reasons, as they were before 1943, from seeking to join the overseas representational services.'[44] The language used is emphatic: 'We would deplore anything tending to narrow the field of recruitment to the Foreign Service.'[45] No documentary evidence exists to illustrate what the FSA thought of the Committee decision but the FSWA was jubilant, as a digest from the British press from February 1964 reproduced in the newsletter and illustrated by an FSWA member testifies.[46]

This euphoria had much to do with the report's recognition of the wives as an emerging force – it praised their 'excellent work in many fields' – but more simply the wives were happy not to have to wait so long to see their children.[47] At their AGM in May 1964 the chair, Mrs Wilson, said that 'families have had the thrill of an Easter holiday spent together and the happy feeling that there are only three months to go till the children come out again for a second time ...'[48] However, one diplomatic wife's retrospective comments support Cromwell's assertion that the Diplomatic Service was more likely to demand conformity from its new recruits than effect any significant change.[49] In a private letter written a decade after the Plowden Report was published, she wrote:

> Loyalty, in the form of responsibility with no power appears to be demanded of wives; the Service should begin to consider the loyalty due

to a wife from her husband, to children by their father, to the united family by the husband's employer and should seriously try to place all these relationships within a modern framework. So much that the DSWA has done has accepted this framework ... I forgot to tell you that after the Plowden report I was dining with the Plowdens and Edwin [Lord Plowden] told me that he had been surprised at the modesty of the wives' demands: he thought in fact that they would have demanded far more.[50]

The importance of the Plowden Report and its timing for the Foreign Office family cannot be underestimated. The intimate tone of the evidence given by individual FSWA and FSA members points to an understanding and receptiveness towards their children's experiences; their frank articulation of grievances to the Plowden Committee provoked the Foreign Office into giving greater thought to its families and to the financial difficulties experienced by the post-1943 recruits. The Diplomatic Service family, although necessarily bound by the demands of a diplomat's job (at this point, this role was exclusively filled by the father), saw attitudes change following Plowden. In conventional narratives of the Foreign Office, Plowden's significance is the merger and creation of the Diplomatic Service, but by looking at the report through a different lens a new story about its impact on families emerges and the impression of success among Foreign Office wives gave them confidence to go on to mount further campaigns in coming decades.

Youth culture, counter-culture, third culture. Diplomatic Service families during the 'swinging sixties'

According to Harrison 'The phrase "the sixties" conjures up at least four images: of youth in revolt, relaxed manners, political radicalism and puritanism repudiated ...'[51] A decade closely associated with the young, its vibrancy and upheaval were closely linked, in the first instance, to what Green has described as the 'freshly minted social group' – the teenager – and a number of opportunities with the potential to cause friction between generations were available from the early stages of the period covered in this chapter. A good example is the Campaign for Nuclear Disarmament (CND), launched in 1958, which offered young people an alternative affiliation to the often military structures of existing movements; 'its mood chimed in with the erosion of the self-disciplined hierarchies of war and empire, whose structured ideals now retreated before a new romanticism that was more egalitarian and participatory in nature.'[52] Subcultures and trends attractive to young people also formed around music and fashion. Olivia Tate*, at boarding school in the 1950s, recalled that she and the other girls:

> weren't particularly restricted at all ... we used to have all the pop music, we all went away at holidays and weekends and came back with all our forty fives to play on the gramophone ... we probably had a broader base

than a lot of the English kids because all the Caribbean girls would bring over music ... and also get music from America ... not just Elvis but all sorts of things so I think we must have had a broader base than many English schools ...[53]

However, many youth subcultures, like the well-known Teddy Boys and Mods and Rockers, attracted the young working class who were already employed, had money to spend and freedom to go out in the evenings and at the weekends. Along with other middle class children, expected to remain in education for longer, Diplomatic Service children – often detained in boarding schools for long periods – were denied the freedom and the means of identifying with groups like these. Indeed, evidence collected for this book indicated very strongly that although some diplomats' children were aware and in favour of societal changes – political and recreational – they were not apt to follow the path of 1960s rebellion. Scholars in this field and former Diplomatic Service children who provided interview evidence for this project suggested possible reasons. Coles wrote that 'the FCO and its members seemed to be about fifteen years out of date, compared with the rest of the country' and this statement certainly seems accurate with regard to its response to the 'swinging sixties'.[54] Andrew Graham, a diplomat's son born in 1956, who enjoyed a distinguished military career as a high ranking army officer, recalled an au pair who accompanied the family to Kuwait in 1966. 'Elizabeth who appeared in a mini-skirt and brought the whole Corniche of Kuwait to an absolute grinding halt and Dad had to say "I think it's a bit ... very fashionable in England but it's not quite ..."'[55] The Graham family's reaction to the mini-skirt, a garment which, according to Sandbrook, 'completed the stereotypical look of the Swinging Sixties', is interesting because it indicates polite disapproval and an unwillingness to explicitly express exactly what is wrong with wearing such a skirt on Kuwait's Corniche. Not only did Andrew Graham's father feel it necessary to speak to the au pair but, from the tone of the interview, we guess that Andrew Graham himself feels that his reaction was appropriate. The Grahams' reaction was no doubt due in part to the demands of cultural sensitivity in a Muslim country, but there is also a real sense that wearing a mini-skirt is not a correct form of dress for a young woman connected to a diplomat's family (or for a young woman, full stop) and Andrew Graham appears to support his father's instinct to maintain the social standards of the past. Harrison has commented on the relaxation of correct uniform and dress during the 1960s. But correct forms of dress were of great significance to the Diplomatic Service and to other British institutions with a ceremonial role.[56] One diplomatic wife recalled an experience at (an unnamed) post:

> I can remember being rung up by the ambassador's wife's secretary and being rebuked for being seen out without a hat or was it coming to lunch without a hat? Yes the latter. So I bought a small hat which cost a lot of money and wore it, but as she was very small and I tall, I heard that she never saw it and I remained on the black list.[57]

For some children of diplomats, parental influence was particularly strong. One contributor said that rebellion was something his father simply would not have countenanced. Eleanor King★, born in 1954, also cited her parents' influence when she recalled being 'totally cut off' from popular culture of any kind and that her parents were keen for her to 'share their contempt' towards it. At post with her parents in Chile during the 1960s she got into conversation with an embassy neighbour who asked her:

> 'Do you like the Beatles or the Stones?' and I hadn't the faintest idea, I'd never heard of them. And she said 'Well … The Stones are like the bad boys, they take drugs and they're really bad and the Beatles are good.' Basically… so I remember saying 'Oh, I like the Beatles, then.'

King★ also reflected on another, more practical, reason why Diplomatic Service children might not have immersed themselves in the culture of the 'swinging sixties'. Wherever they were in the world, Diplomatic Service families occupied a privileged position, living in exclusive neighbourhoods or protected institutions. When she made her choice between the Beatles and the Rolling Stones Eleanor King was on holiday from a girls' Catholic boarding school that she attended in the UK.

> I had … this sense of a world being out there that I couldn't get at, and knew it was to do with the strange … you know, the way we weren't really living in this place. But school was just an institution, there you are in Berkshire and you had to wear the headscarf and you were controlled the whole time …[58]

Both King★ and another contributor who spent the 1960s in Catholic schools felt that they finally 'found life' when they arrived at Oxford. A third correspondent who attended progressive boarding school during the 1960s was permitted slightly more freedom, recalling that she and her peers were influenced by hippie culture, wore long dresses and jeans and listened to pop and rock music. Nonetheless her observations on boarding schools as a protected place, set aside from the world at large, coincide with those of King★: despite her school being co-educational she did not go on a proper date until she was at university; all life was lived within the school boundaries.

Unsurprisingly then, in answer to a specific interview question 'Did you rebel as a teenager?', contributors to this book who were young in the 1960s all stated that they had been very well behaved and never rebelled.[59] Some expressed the opinion that their background was not sufficiently settled to engender or support rebellious feelings. Reluctance to rebel was also related to family separation; some diplomatic children considered that they had not rebelled against their parents because they did not live with them or know them very well. They also pointed out that separation in the form of boarding school meant that they did not have the same social relationships as other young people and felt that this influenced whether or not they felt able to rebel. Diplomatic Service children appeared not rebel because they did not have a steady enough platform from which to do so. They did not really know their parents, but they did not feel entirely comfortable with their friends at school either. Some participants, however, could be called if not 'naturally rebellious', then apt to challenge the status quo in a way that others did not. Eleanor King* (b. 1954) remembered feeling uncomfortable early on with the structure of the diplomatic households in which she lived as a young person:

> I was very much sort of 'What right have we to have servants anyway?' These people are in their own country. You know, I reacted very badly to the caste superiority as far as I could see it as a child. And then when I got to the convent, you know, the way the girls talked ... I evolved the politics while knowing nothing about politics, I evolved the politics out of just not liking what I was seeing and hearing. But I really hated, you know, the authority of having to be British. I can remember when Churchill died and somebody, a journalist, came to the door in Chile, and I think it was just me doing homework somewhere and he said 'And so, how are your parents marking this?' and I said, innocently, but actually knowing that it was not the right thing to say, 'Oh, my Mum's just gone riding as usual.' And feeling you know, that I was dealing my little blow there against ... against representation.[60]

However, evidence does suggest that diplomatic parents, enduring separation from their children, were very concerned by the lurid stories of 1960s' excess that attracted a great deal of publicity, and that they were especially worried about drug use. A memo written on behalf of the FCO Welfare Section's Medical Adviser raised awareness of the difficulties experienced by young people aged between eighteen and twenty-one whose parents were overseas and questioned why there was not one concessionary fare per year paid for all young people from Diplomatic Service families rather than just for those at university. Notably, the report concluded that, in contrast to the high expectations of independence and resilience from Diplomatic Service children discussed in the first chapter, children 'educated in boarding schools and going to see parents in overseas posts for the holidays' were 'very sheltered' and their problems linked to 'immaturity and inexperience'. The report went on that 'they obtain far more personal freedom and independence than they have ever known and are exposed to the stresses of a different type of work and today's permissive society'. 'Anxiety states, schizophrenia and drug addiction commonly begin in this age group' is the alarming conclusion.[61] However, despite the fact that boarding schools (and their occupants) were largely isolated there remained scope for vices. Millham and Lambert's study of boarding school life *The Hothouse Society*, based on evidence collected from boarding school pupils in the mid-1960s and published in 1967, demonstrated that drugs were readily available in boarding schools: 'We have come across ... small drug circles in famous public schools, progressive and integrated schools.'[62] This was not the only dubious habit: in the course of their research, Millham and Lambert also noted the prevalence of smoking, theft (from fellow students and in shops) and gambling.

It is also worth noting that newspaper stories which detailed the 'permissive society' during this period often used the 'diplomat's daughter' as an exemplar of a staid, old-fashioned world, especially as the 1960s progressed. In 1969 the *Mirror* newspaper ran two separate stories concerning diplomats' daughters. February saw a small piece about a diplomat's daughter who 'stripped off' on a roof top in New York to advertise a novel called 'The Voyeur'.[63] In December, Serge Gainsbourg's film *Cannabis* featured British actress Jane Birkin as a 'pot smoking orgy girl' who is the daughter of the British Ambassador in Paris.[64]

The spectacular exception to prove this obedient rule is Teresa Hayter, daughter of Sir William Hayter, a diplomat who entered the Foreign Office in 1930 and who held posts in Paris and Moscow. Born in China in 1940, Teresa Hayter was among the last cohort of debutantes presented at court; with three other young women she was 'known in the gossip columns ... as "the blue-stocking debs"' because she was assured a university place at Oxford in the same year.[65] In the early 1970s Hayter turned her back on Oxford education and has been quoted as saying 'The function of Oxford Education is to train people to service and perpetuate capitalism'.[66] She embraced revolutionary Marxism: 'When the revolution comes I shall be helping man the barricades with guns and petrol bombs.'[67] Hayter wrote two books in 1971 – an auto-biography, *Hayter of the Bourgeoisie*, and *Aid as Imperialism,* a critique of western aid practices – and ran a radical bookshop in Leeds. She was described as 'dull, silly, incompetent and irrelevant' and – rather like the example of Joe Strummer in the introduction – criticised for espousing radical left-wing politics after the perceived privilege of her upbringing, but she did not renounce her political beliefs. Now in her eighties, she remains an activist and outspoken critic of the asylum system. Nonetheless Hayter's early travels – to China, the US and France – as the child of a diplomat do not appear to have contributed to her radicalism in the same way that later travels, after university, did.

The consensus among historians of the late twentieth century, however, is that the radical counter-culture that typifies the 1960s in the public imagination was short-lived in Britain. It appears, that, in comparison with counterparts in France, Germany and the US, very few young Britons adopted 'permissive' lifestyles or engaged in political protests.[68]

> Notions of sixties counter-cultural movements as agencies of opposition with the power to fundamentally restructure society seem rather exaggerated. Throughout the sixties these movements never attracted more than a small minority of the young, contemporary research testifying time and time again to the 'conventionality' of most youngsters.[69]

But during this period a comparable alternative 'culture' that would prove to be far more relevant and applicable to Diplomatic Service families was on its way to formulation and recognition. The term 'third culture' was conceived by US anthropologists Ruth Hill Useem and John Useem in 1958 to depict the phenomenon they had observed amongst the Americans living and working in India, whose behaviour they had been commissioned to document. The Useems observed that the American and the Indian nationals maintained their own cultures when interacting within those groups, but that different methods of interaction and behaviour were exhibited when the two groups were brought together, thus creating what they termed a 'third culture'. Ruth Hill Useem began to observe the behaviour among the women and children of American families in India, her own children included, and defined the children as 'third culture kids'. Since the 'third culture' was defined in 1963, awareness of 'Third Culture Kid' (or TCK) discourse has gathered pace and has been a subject of lively discussion: a 'handbook' for 'TCKs' was published in 1999 and a well-maintained website echoes the views and circumstances of its community.[70] The current definition is as follows:

> The TCK is a person who has spent a significant part of his or her developmental years outside the parents' culture. The TCK frequently builds relationships to all of the cultures, while not having full ownership in any. Although elements from each culture may be assimilated into the TCK's life experience the sense of belonging is in relationship to others of similar background.[71]

As will be seen, many aspects of the experience of children in the British Diplomatic Service resonated with the TCK phenomenon, yet the terminology of Third Culture Kids itself was not widely adopted by contributors to this study, nor by FCO authorities. The reasons for this are very complex but, in short, the 'TCK' is more frequently associated with various aspects of US culture and with international schools, which were infrequently used by British diplomats during the period covered.

The third 'concessionary' journey

The Plowden Committee did not feel able to grant a third concessionary return journey for children of Diplomatic Service families, due, in part, to financial constraints and to an awareness of external opinion: from within the civil service departments and outside government. However, the Foreign Service Association, renamed as the Diplomatic Service Association (DSA), in line with the Report's recommendations for an amalgamated service, which had represented so strongly on behalf of family life to the Plowden Committee, did not lose sight of this goal. In 1970, as the new decade began, discussions on the third concessionary journey were held at the very highest level of the Foreign Office and in Whitehall. The third journey was eventually granted at the end of 1971.[72]

Backers of the campaign for the third fare built on the reasons that had been submitted to the Plowden Committee to justify requests for subsidised journeys every school holiday. They stressed again the trials of family separation, damage to morale and the dilemmas of providing accommodation for children's holidays, either through private holiday homes or with extended family. In March 1971 the DSA submitted a paper to its managing committee, recommending a further concessionary journey. The DSA recalled that it 'represented strongly to the Plowden Committee that children at boarding school in the United Kingdom should be able to join their parents at public expense three times a year'.[73] This paper stated that diplomats and their wives were not able to concentrate efficiently on their work while they were worried about their children (notice that wives were supposed to work too) and these concerns were exacerbated by 'further radical social changes' since the publication of Plowden in 1964. The report's author wrote: 'The "generation gap" has grown wider as young people have been increasingly exposed to problems quite different in degree and kind from those experienced by their parents.' Although interviews proved that teenage rebellion among diplomats' children was almost non-existent – as we saw above – diplomatic parents voiced increasing anxieties about older children and young adults during the 1960s and 1970s. A confidential paper produced by the Diplomatic Services Wives Association (DSWA) refers euphemistically to the 'influences and stimuli to which their children are subject ...' and worries that 'These influences and stimuli are particularly relevant in the case of young adults'.[74] There was no doubt that parents were particularly concerned about drug use, which by the late 1960s was a recognised part of boarding school life, in line with developments in national youth culture. But drug use in boarding schools was not nearly as widespread as absent parents suspected. This was largely due to boarding schools' physical isolation and their very structured schedules.[75] Diplomatic Service parents nevertheless felt that more contact with their older children would enable them to exert greater levels of control and supervision.

The DSA paper went on to assert that many FCO employees were still experiencing financial hardship, despite the Plowden Report's best efforts to provide a remedy. The disparity between officers serving at distant posts and those who spent large parts of their career in Europe, from which fares were not financially prohibitive, were often pointed out and the spectre of the private income led to further inequalities. Junior grade members of staff, as well, were at a disadvantage as their lower salaries made it difficult for them to afford air fares. One wife wrote bitterly to the DSWA:

> a private income, luck in postings or a certain seniority within the Service enable an officer to buy his way out of this dilemma ... It does not make for fellow feeling in a post where some mothers but not others can 'buy' the company of their children.[76]

The contrasts between benefits provided to diplomatic staff and those received by the employees of private companies with offices overseas also gave cause for disquiet and invited comparisons. 'The great majority of firms give terms at least as good as does the Diplomatic Service,' the paper stated, 'and a growing number of firms, particularly banks, ensure that parents and children are united every holiday.'[77] This suggests that the Plowden Report's decision to unite the Foreign Service, Commonwealth Service and Trade Commission Service in order to better address 'the problem of earning our [i.e. Britain's] living in the world' led to greater familiarity between diplomats and private sector workers in expatriate communities, resulting in inevitable comparisons of lifestyle and benefits.[78] However, as many of the essays compiled by Coles and Fechter make clear, expatriate Britons had always formed tight-knit communities overseas and conditions of service were no doubt regularly discussed.[79]

The DSA set out this careful case in response to a short background brief for the April 1971 meeting of the Diplomatic Service Whitley Council.[80] When the DSA had announced at a prior meeting of the Council that it intended to push for a third journey, 'It was emphasised ... that the case would need to be a good one and that moral indignation was not enough'.[81] In the event, however, the Council's fear that moral outrage would be used to form an argument rather than the presentation of hard facts was not justified. It was found that support for the third journey came from many quarters, in response to a circular sent out to all posts to canvas opinion.[82] Derek Tonkin, Head of Chancery in Wellington, New Zealand, made the case for the inequality between staff based in Europe and those far further afield. In a letter entitled 'The Loneliness of the Long Distance Officer', Tonkin set out the views of his own staff who had suggested that the Civil Service Department should establish a touring group with the sole responsibility of investigating the problems of distant posts. Tonkin wrote:

> Apart from making arrangements for the 'third' holiday, there are the usual run of teenage, house purchase and ageing relatives and other problems where in the last resort an appeal on compassionate grounds would possibly produce financial authority for a special journey; the point being, however, that matters ought never to reach such proportions, and indeed do not for those fortunate officers living within 3000 miles of London.[83]

An impassioned letter from an anonymous wife in Accra (unkindly described by an FCO official in London as 'a bleat') adopted a decidedly emotional tone, although not without expressing a valid argument. Her letter acted as a reminder that British society and its attitude towards the family were changing and referred directly to local government policy towards children in the UK and to their cultural representation:

> It is vital to stress the rights and interests of the child. Since the war there has been a revolution in thinking about the child's need for his parents and a secure home. I believe that current practice in Local Authority Children's departments is to reunite mother and child, child and family as often and as quickly as possible ... The documentary 'Cathy Come Home' made

its impact by showing Cathy's children torn from her by the authorities. This is what FCO mothers have to put up with and I am sure that if a Local Authority treated families in this way there would be a public outcry.[84]

In the case of the third journey, it transpired that comparisons which had been made with private companies had not been as influential as FCO campaigners originally thought they might be. Ultimately the FCO's Administration were typically moved by the need to:

administer the service equitably and consistently all over the world, it was impossible to send a man with a family to the other side of the world without putting him at a disadvantage with those serving nearer home. The desire to remove this inequality was the main argument in the exercise.[85]

Despite these justifications, there is evidence that very senior civil servants were willing to act in what they perceived to be the children's interests. Permanent Under Secretary for Foreign Affairs (PUS) Denis Greenhill had already noted at the bottom of an official memo:

I had an opportunity of talking to the PM and Lord Jellicoe about this recently. They both seemed well disposed to the idea. The minute should also go to the Chancellor of the Exchequer. The strongest justification is the welfare of the children.[86]

The third journey was eventually granted late in 1971. Described as a 'long and tough slog', the successful addition of this much sought after allowance was not reported with the same unbridled enthusiasm by the DSWA newsletter as Plowden had been seven years earlier.[87] However, it is noticeable, as the DSWA newsletters progress, that attitudes towards allowances underwent a subtle course of change. Civil servants and their wives gave evidence to the Plowden Committee in the hope of being able to counteract the threat of real financial hardship; but by the time the third fare was granted, allowances were viewed by FCO employees with far more of a sense of general entitlement, with the FCO itself seen as the great provider, to be blamed if circumstances were not ideal. For example, a new concern presented itself to FCO families when the age of majority was lowered from twenty-one to eighteen as part of the Family Law Reform Act of 1969. A letter published in the first DSWA newsletter of 1971 expresses dissatisfaction that the FCO did not provide allowances for children over eighteen and suggests that it is to some extent the cause of a young woman's 'abandonment': 'In my own case I have an 18 year old daughter who, at the end of her last school term was, so far as the office was concerned, simply abandoned on the school doorstep to fend for herself …'[88] As Coles has put it, the system of allowances could lead to '"a tender trap" in which families could find themselves bureaucratically enmeshed'.[89]

Conclusion

This chapter has seen that, on the whole, the radical social reforms of the 1960s that were hugely influential in European and North American society touched the culture of Diplomatic Service life only tangentially and could even be said to emphasise the anachronistic tendencies of its systems and personnel. It was not until the Plowden Report, published in 1964, that issues specific to the diplomatic family – of long periods of separation and the expense of boarding schools – were directly addressed. Plowden's recommendations began a new era for the Diplomatic Service family, which, though it retained its traditional shape through long-held practice in a habitually conservative organisation, began to grow in confidence under the influence of social changes both inside and outside the Foreign Office. These changes, however, did little to galvanise the spirits of Diplomatic Service youth, none of whom took part in any significant acts of rebellion. Some thoughtful contributors expressed dissatisfaction with Diplomatic Service life – especially at post – and wider social and political practices. This impression of dissatisfaction can be glimpsed through evidence submitted to the Plowden Report by the Diplomatic Service Wives Association and the Diplomatic Service Association, both of which were formed in the early 1960s: the Foreign Office Administration was forced to face the reality that wives and children would not passively accept the many demands and privations imposed by Diplomatic Service life. The campaign for the third 'concessionary' journey, conducted less than a decade later, demonstrated a newfound confidence and solidarity among groups like the DSWA and ever greater anxieties about rapidly changing social standards.

Notes

1 See Hunter Davies, *The Beatles* (London: Ebury, 2009), 298–299.
2 Ibid., 300.
3 Ibid.
4 Sally James, *Diplomatic Moves: Life in the Foreign Service* (London: Radcliffe, 1995), 13.
5 Beryl Smedley, *Partners in Diplomacy* (Ferring, West Sussex: Harley Press, 1990), 156, quotes an anonymous diplomatic wife: 'In 1955 when our eldest daughter had to be left behind at boarding school it was pre-Plowden …'
6 Valerie Cromwell, 'A World Apart: Gentleman Amateurs To Professional Generalists', in *Diplomacy and World Power Studies in British Foreign Policy 1890–1950*, ed. Michael Dockrill and Brian McKercher (Cambridge: Cambridge University Press, 1996), 8.
7 Brian Harrison, *Seeking a Role: British Society 1951–1970* (Oxford: Clarendon Press, 2009), 50, provides a useful introduction to the post war Labour government's education policy and the 'educational landscape' of the UK.
8 Ibid., 194.
9 Plowden Report, iv.
10 Committee on Representational Services Overseas [hereafter RSO] (C) (62) 1st meeting 12.11.62 Part 2 (c).
11 RSO (C) (62) 1st meeting 12.11.62 Part 2 (c).
12 'Minutes of the Annual General Meeting', *FSWA Newsletter,* 3 July 1961, 1.

13 Masha Williams, 'Foreign Service Wives' Association: How It Started', *DSWA Newsletter* Autumn 1985, 52–54; RSO (62) 54.

14 Williams, 'How It Started', 52–54.

15 'Minutes of the Annual General Meeting', *FSWA Newsletter*, 3 July 1961, 1.

16 Katie Walsh, 'Travelling Together? Work, Intimacy, and Home amongst British Expatriate Couples', in Anne Coles and Anne-Meike Fechter, *Gender and Family among Transnational Professionals* (London: Routledge, 2008), 64.

17 The Foreign Service Association was the forerunner of the Diplomatic Service Association trade union which represents 'policy entrants' or 'fast-streamers'; it was formed in 1960 at around the same time as the Foreign Service Wives Association. The evidence submitted to the Plowden Committee by the FSA can be found at RSO (62) 47.

18 Katie Hickman, *Daughters of Britannia: The Life and Times of Diplomatic Wives* (London: Flamingo, 1999), 64.

19 RSO (62) 54.

20 Ibid.

21 Ibid.

22 RSO (62) Sub Committee on Terms of Service 12/11/1962.

23 Harry Hendrick, *Children, Childhood and English Society 1880–1990* (Cambridge University Press, 1997), 33.

24 RSO (62) 54.

25 RSO (62) 47.

26 Zara Steiner, 'The Foreign and Commonwealth Office: Resistance and Adaptation to Changing Times', in *The Foreign Office and British Diplomacy in the Twentieth Century*, ed. Gaynor Johnson (London: Routledge, 2005), 25.

27 FSO (62) 47 FSA Evidence to Committee.

28 Nonetheless, a sense of outrage was often eventually overcome. The 'process of acculturisation' that Steiner describes in 'Resistance and Adaptation to Changing Times' (23) clearly influenced many diplomats to re-think their divided loyalty between FO and family.

29 Laura King, *Family Men* (Oxford: Oxford University Press, 2015), 16–49.

30 King, *Family Men*, 89–122.

31 RSO (62) 47.

32 Ibid.

33 In the years before the Second World War psychologists of child development affected a shift from the belief that juvenile delinquency was a result of environment – especially industrial urban environments – and was instead attributed to individual psychology. For a good discussion of 'the problem child', its pan-European significance and proposed solutions, see Hugh Cunningham, *The Invention of Childhood* (London: BBC Books, 2006), 178; John Stewart, *Child Guidance in Britain 1918–1955* (London: Pickering and Chatto, 2013), 175. The 'problem family' also became a cause for concern in the 1940s.

34 Hickman, *Daughters of Britannia*, 224. As the adult 'children' described by Hickman's 'Spokesperson' were in their thirties and forties they would have been children in the 1950s and 1960s, some possibly coinciding with Plowden. Ruth Dudley Edwards' *True Brits: Inside the Foreign Office* (London: BBC, 1994), 219, took the 'problem child' appellation still further when she wrote that Diplomatic Service children have a high incidence of suicide. Participants did report incidents of unhappiness but there is not sufficient evidence to support this claim.

35 RSO (C) (62) 2nd meeting 22 November 1962.

36 RSO (C)(63) 1st Meeting 14 January 1963.

37 RSO (C) (63) 3rd Meeting 19 April 1963.

38 RSO (62) 54.

39 RSO 63 (C) 3rd Meeting 19/04/1963.

40 Plowden Report, 148.
41 Ibid.
42 Ibid., 122.
43 Plowden Report, 120.
44 Plowden Report, 119.
45 Ibid.
46 'The Press on Plowden or Enter the Whitehall Whizz-Kids', FSWA newsletter July 1964. Interestingly the 'schizophrenic' attitude shown by diplomatic families towards privilege and hardship is illustrated in this piece by a short, guilty sentence: 'Dare we admit, after all that, that *some* of them can be fun.' This is in response to an article from the *Sunday Times* that stresses the onerous qualities of 'a fearful round of diplomatic cocktail parties'.
47 Plowden Report, 129.
48 FSWA Newsletter, January 1964, 3.
49 Cromwell, 'A world apart', 8–9, criticises the Foreign Office decision to produce the guidance/etiquette booklet written by Marcus Cheke: implying that, given the opportunity to refresh out of date practices it assumed instead that these would continue and new recruits would conform to them.
50 Diana Richmond to Mrs Wilford (on the DSWA Committee), 31 October 1971. Smedley Box B.
51 Harrison, *Seeking a Role*, 472.
52 Ibid., 483.
53 Olivia Tate★. Interview by Author. 10 March 2014.
54 Ann Coles, 'Making Multiple Migrations: The Life of British Diplomatic Families Overseas', in Coles and Fechter, *Gender and Family*, 143.
55 Andrew Graham. Interview by author. 21 May 2015
56 Harrison, *Finding a Role*, 494.
57 Diplomatic wife letter to Beryl Smedley in response to a request for material for Smedley's book *Partners in Diplomacy* undated Smedley Box B.
58 Interview. Eleanor King★. 11 July 2014.
59 This absence of rebellious spirit was also true of participants born earlier and later, whose youth did not coincide with the 1960s.
60 Author interview. Eleanor King★. 11 July 2014.
61 TNA XEG5/22 Possible introduction of third concessionary journey for children of Diplomatic Service personnel; travel from UK schools to overseas postings; 1970 Jan 01–1971 Dec 31; Dr Ruth Lloyd-Thomas, Welfare Section 29 July 1971.
62 Royston Lambert with Spencer Millham, *The Hothouse Society. An Exploration of Boarding School through the Boys' and Girls' Own Writing* (Harmondsworth: Penguin, 1974), 279.
63 'Sarah strips off above New York', *Daily Mirror*, 7 February 1969.
64 'The private life of diplomat's daughter', *Sunday Mirror*, 7 December 1969.
65 Fiona MacCarthy, *Last Curtsey: The End of the English Debutante* (London: Faber and Faber, 2006), 75.
66 Teresa Hayter quoted in Melvin Jonah Lasky, *On the Barricades, and Off* (New Brunswick, Oxford: Transaction, c.1989), 59.
67 Ibid., 58.
68 See Dominic Sandbrook, *White Heat. A History of Britain in the Swinging Sixties* (London: Little, Brown, 2006), 491–515.
69 Bill Osgerby, *Youth in Britain since 1945* (Oxford: Blackwell, 1998), 91.
70 The article that contained the now regularly quoted definition of 'the third Culture' was John Useem et al., 'Men in the Middle of the Third Culture: The Roles of American and Non-Western People in Cross-Cultural Administration', *Human Organisation* 22:3 (1963): 169–179. David C. Pollock and Ruth E. Van Reken,

Third Culture Kids Growing Up Among Worlds (London: Nicholas Brealey Publishing, 2009), reads as part history, part theory and part self-help guide.

71 Pollock and Van Reken, *Third Culture Kids*, 20–23.
72 TNA files FCO 77/204 and FCO 77/205 contain a useful and comprehensive set of papers that describe the process by which the Diplomatic Service Association lobbied for the third concessionary journey and on which this section is largely based. However, the DSA did work in tandem with the Foreign Service Wives Association which had become the Diplomatic Service Wives Association or DSWA and with Medical and Welfare departments within the Foreign Office. Again, this is a good example of the 'patchy' nature of surviving evidence surrounding family matters; and the 'patchy' approach to them within the FCO itself.
73 TNA, FCO 77/205.
74 Confidential paper by the DSWA, 'The DSWA's view of various aspects of diplomatic life', 13 May 1976. Smedley Box B.
75 Lambert and Millham, *Hothouse Society*, 279–281.
76 TNA FCO 77/205 Possible introduction of third concessionary journey for children of Diplomatic Service personnel; travel from UK schools to overseas postings; 1970 Jan 01–1971 Dec 31; Mrs Norah Reid 6 June 1971.
77 TNA, FCO 77/205.
78 Plowden Report, 5.
79 Ibid., 3.
80 A Whitley Council is a body made up of officials and trade unions which exists to negotiate conditions of service. See David Summerhayes, 'The Staff Side – What's That?' *DSWA Newsletter*, Autumn 1972, 51–52.
81 TNA, FCO 77/205.
82 TNA files do not appear to contain a copy of this circular but reference is made to it in much of the correspondence held on files FCO 77/204 and FCO 77/205. It appears to have been prompted by the DSWA.
83 TNA, FCO 77/205.
84 TNA, FCO 77/205.
85 Ibid.
86 Ibid.
87 'Report of DSWA AGM 17/05/1972', *DSWA Newsletter*, Autumn 1972, 7.
88 'Letters to the Editor from Mrs Norah Reid, Accra', *DSWA Newsletter*, Spring 1970, 12.
89 Coles, 'Making Multiple Migrations', 129.

Bibliography

Beryl Smedley Archive Box A: Questionnaire responses from a survey of diplomatic wives.

Beryl Smedley Archive Box B: Press cuttings and FSWA/DSWA material relating to diplomatic wives.

Coles, A. and Fechter, A. eds. *Gender and Family Among Transnational Professionals* (London: Routledge, 2008).

Cromwell, V. 'A World Apart: Gentleman Amateurs to Professional Generalists', in *Diplomacy and World Power Studies in British Foreign Policy 1890–1950*, ed. M. Dockrill and B. McKercher (Cambridge: Cambridge University Press, 1996).

Cunningham, H. *The Invention of Childhood* (London: BBC Books, 2006).

Daily Mirror, 'Sarah strips off above New York', 7 February1969.

Davies, H. *The Beatles* (London: Ebury, 2009)

DSWA Newsletter Spring 1970; Autumn 1972; Autumn 1985.

Dudley Edwards, R. *True Brits: Inside the Foreign Office* (London: BBC, 1994)

Evidence submitted to the Committee on Representational Services Overseas (The Plowden Committee). Bound volumes: Committee on Representational Services Overseas: Evidence RSO (62) 1–30.

Evidence submitted to the Committee on Representational Services Overseas (The Plowden Committee). Bound volumes: Committee on Representational Services Overseas: Evidence RSO (62) 31–59.

Evidence submitted to the Committee on Representational Services Overseas (The Plowden Committee). Bound volumes: Committee on Representational Services Overseas: Evidence RSO (63) 1–12.

Evidence submitted to the Committee on Representational Services Overseas (The Plowden Committee). Bound volumes: Committee on Representational Services Overseas: Evidence RSO (63) 13–59.

FCO Files 77/204; 77/205 *The National Archives, Kew.*

*FSWA Newsletter*July 1961, July1964.

Harrison, B. *Seeking a Role: The United Kingdom 1951–1970* (Oxford: Clarendon, 2009).

Harrison, B. *Finding a Role: The United Kingdom 1970–1990* (Oxford: Clarendon, 2011).

Hendrick, H. *Children, Childhood and English Society 1880–1990* (Cambridge: Cambridge University Press, 1997).

Hickman, K. *Daughters of Britannia: The Lives and Times of Diplomatic Wives* (London: Flamingo, 1999).

James, S. *Diplomatic Moves: Life in the Foreign Service* (London: Radcliffe Press, 1995)

King, L. *Family Men* (Oxford: Oxford University Press, 2015).

Lambert, R. with Millham, S. *The Hothouse Society. An Exploration of Boarding School Life Through the Boys' and Girls' Own Writings* (London: Pelican, 1974).

Lasky, M.J. *On the Barricades, and off* (New Brunswick, Oxford: Transaction, c.1989)

MacCarthy, F. *Last Curtsey: The End of the English Debutante* (London: Faber and Faber, 2006).

Minutes of the Committee on Representational Services Overseas (The Plowden Committee). Bound Volumes: Committee on Representational Services Overseas: Minutes Main Committee.

Minutes of the Committee on Representational Services Overseas (The Plowden Committee). Bound Volumes: Committee on Representational Services Overseas: Minutes Sub Committee.

Osgerby, B. *Youth in Britain since 1945* (London: Blackwell, 1998).

Pollock, D. and Van Reken, R. *Third Culture Kids Growing Up Among Worlds* (London: Nicholas Brealey Publishing, 2009).

Report of the Committee on Representational Services Overseas appointed by the Prime Minister under the Chairmanship of Lord Plowden, 1962–63 (Representational Services Overseas), 1963–1964 Cmnd. 2276.

Sandbrook, D. *White Heat: A History of Britain in the Swinging Sixties* (London: Little, Brown, 2006)

Smedley, B. *Partners in Diplomacy* (Ferring, West Sussex: Harley Press, 1990).

Steiner, Z. 'The Foreign and Commonwealth Office: Resistance and Adaptation to Changing Times', *Contemporary British History* 18:3 (2004): 13–30.

Steiner, Z. 'The Foreign and Commonwealth Office: Resistance and Adaptation to Changing Times', in *The Foreign Office and British Diplomacy in the Twentieth Century*, ed. G. Johnson (London: Routledge, 2005).

Stewart, J. *Child Guidance in Britain 1918–1955: The Dangerous Age of Childhood* (London: Pickering and Chatto, 2013).

Sunday Mirror 'The private life of diplomat's daughter', 7 December1969.

Useem, J., Useem, R. and Donoghue, J. 'Men in the Middle of the Third Culture: The Roles of American and Non-Western People in Cross-Cultural Administration', *Human Organisation* 22:3 (1963): 169–179.

Walsh, K. 'Travelling Together? Work, Intimacy, and Home amongst British Expatriate Couples in Dubai', in *Gender and Family Among Transnational Professionals*, ed. A. Coles and A. Fechter (London: Routledge, 2008).

3   1972–1985

In 1985 Foreign Office society came under the scrutiny of two separate inquiries. The first was an internal document: that year marked the thirtieth anniversary of the Commonwealth Relations Office Wives Society (the CROWS), the twenty-fifth anniversary of the Foreign Service Wives Association (the FSWA) and the twentieth anniversary of the Diplomatic Service Wives Association (DSWA) which was formed to amalgamate the preceding two after the recommendations of the Plowden Committee united the Foreign Service and the Commonwealth Relations Office in 1965. The anniversary edition began with three articles which considered the origins of the wives' associations and the continuing issues they faced, 'by three people involved in the formative years of the CROWS, the FSWA and the DSWA respectively: Lady Garner, Lady Williams and Lady Crawford'.[1] Many of the issues described – family separation, the demands of a transient lifestyle and recurring re-integration into British society demanded by multiple migrations – were familiar, apparently little changed from the concerns seen in the preceding chapter, expressed to the Plowden Committee in the early 1960s.

The second was external – and the sort of process which had always provoked suspicion among diplomats. Early in 1984 BBC Radio Four broadcast a programme based on the Foreign Office, produced by Anne Sloman, who had already created programmes based on the Home Civil Service and the Treasury. The well-known journalist Simon Jenkins travelled to British embassies in Europe and the US and visited the Foreign Office in Whitehall to conduct interviews and compile details for the broadcast and its accompanying book *With Respect Ambassador.* [2] In the book, Sloman and Jenkins presented a compelling account of the Diplomatic Service in the 1970s, laced with their own, independent, opinions. Among Jenkins' many observations came the report that he felt the Diplomatic Service comprised 'very large numbers of people who were extremely able, who were simply not stretched' and thought some of its well-established practices 'odd': 'the whole thing is done through a sort of prism of classical bourgeois entertaining. I became a little peeved at endlessly being sat between two embassy wives at almost every function ...'[3] However, Jenkins maintained:

DOI: 10.4324/9780429273568-3

I was very impressed by the number of diplomats' wives who were extremely keen to talk to me, in often very critical terms, about the way in which they felt the Service regarded spouses and their problems and we had a great debate as to how far we should use this material in a programme that was essentially about the Foreign Office.[4]

What Jenkins had to say posed two questions. The ambivalent position of Foreign Office 'camp followers', who did so much for the institution but who held no official footing, was the first. The second was that Diplomatic Service wives seemed very eager to speak plainly to Simon Jenkins: they felt that the Foreign Office did not listen to them and they wanted to express their grievances through as many channels as possible.

The two publications outlined above contained discussions of much that was familiar. But the circumstances in which Diplomatic Service families conducted their lives were changing swiftly, affected by developing national and international trends, which in turn prompted unprecedented complications in the lives of diplomats overseas. This chapter, which examines the period between 1972 and 1985, concentrates on three areas of concern to Foreign and Commonwealth Office (FCO) families which are prominent in the 1985 anniversary newsletter and in Jenkins and Sloman's book. The first section looks at the changing position of women in the UK workforce in the light of second wave feminism and statutory reform. In this period the Foreign Office marriage bar was finally dropped, raising the possibility for greater numbers of female diplomats. New graduates joining the Diplomatic Service were more likely to be accompanied by wives they had met at university who were no longer content to put up with the traditional role of a Foreign Office wife. Discussions on the future of wives in the Diplomatic Service animated DSWA newsletters and their position was addressed in two separate influential commentaries by academics. It was also during this period that the voices of Diplomatic Service children were heard in the DSWA newsletters for the first time. As children in the UK were supported in greater freedom of expression and the concept of children's rights began to develop, the second section of this chapter will look at the ways in which Diplomatic Service children's voices began to be heard – partly in response to the 1979 United Nations (UN) Year of the Child – and will continue to examine the experiences of FCO children alongside trends in the care and treatment of children in the UK.[5] Section three moves on to explore the increase in threat faced by Diplomatic Service families overseas during this time frame. Described in the DSWA newsletter as 'kidnappings, hijackings, bombs and cold-blooded murder', transforming international prospects and the rise in terror for political ends led to a marked increase in the threat of and actual violence experienced by diplomatic families worldwide.

Diplomatic wives and feminism

The early 1970s in Britain saw consecutive acts of government legislation – the Equal Pay Act of 1970, the Sex Discrimination Act of 1975 and the Race Relations Act, 1976 – that resulted in longstanding changes to the workplace. The Sex Discrimination Act, which sought to end prejudice against married persons in the field of employment, had meaningful implications for the Diplomatic Service. In anticipation of the Act becoming law the Foreign Office was no longer able to enforce the marriage bar which required women diplomats to resign on marriage and which, as McCarthy observed, 'was becoming increasingly difficult to justify to the outside world'.[6] The Diplomatic Service marriage bar was lifted in 1973, twenty-seven years after that of the Home Civil Service. Not only did this issue 'a fundamental challenge to the gender order which had underpinned British diplomatic communities since the early nineteenth century', it also introduced into Foreign Office society the potential for women diplomats to shake its uneasy foundations by presenting and identifying as wives and mothers for the first time.[7]

Naturally this fundamental change was slow to take effect. The few women who already occupied senior positions within the Service were older, and unmarried and childless. Younger women who had joined the Diplomatic Service during the 1960s had either chosen not to marry or had been forced to resign and join the ranks of the diplomatic wives. Sheila Skinner 'resigned' in 1964 after marrying fellow diplomat David Skinner. When the couple were posted to South Africa the same year, Skinner recalled:

> I went in to see the then Consul General in Johannesburg and said 'Look, here I am I have a certain amount of experience in the Foreign Office … what sort of job is there for me within the office?' And he said 'Wives do not work. Goodbye.' And that was the end of my career.[8]

Conversely, the experiences of Anne Foster★, a diplomat who married and resigned in 1970, provide a cogent illustration of how much the personality and discretion of an individual ambassador to interpret regulations could make a difference to a married woman who wanted to continue in a diplomatic role. Foster★ remembered: 'I went out to Cuba and the ambassador there was very sympathetic and I worked on some research papers for IRD while I was there but not paid and unofficially …'[9] Thus Foster★ was able to keep up with work within her chosen profession but only by offering her considerable skills and talents for free, and presumably without recognition.

The debates surrounding the unpaid and unofficial status of work undertaken by diplomats' wives, both of a professional and more traditional nature, especially at post, rumbled on. They surfaced again in 1976 when the DSWA's Spring newsletter published an article by Margaret Ibbott entitled 'The Role of a Foreign Service Wife – A Personal View'. In this article Ibbott expressed the opinion that diplomatic wives were not bound to the Service in the way that their husbands were because they were not employed by it: 'however much a wife chooses to help her husband she is not, as a simple matter of fact, an employee.'[10] Ibbott stated that she wanted to take a job outside the embassy and while her husband was perfectly happy with this:

> Others connected with the F.C.O. have … applied pressure designed at the very least to make me feel apologetic … Other wives particularly have made it clear where they think all diplomatic wives' duties lie, with the post in particular and with the F.C.O. in general and that personal interests, professional or otherwise, must come last.[11]

Ibbott's comments about professional interests recalled a case reported to the DSWA four years earlier, in 1972. A confidential note dealt with the case of a young diplomatic wife, a medical doctor, who had asked the FCO whether she could apply for the vacant role of medical adviser in New Delhi, where her husband had been posted. The query was passed to the Civil Service Department Medical Adviser who thought:

> such an appointment could lead to considerable difficulties. She pointed out that many wives might not wish their illnesses known to another wife on the staff, even if she was a doctor. Senior wives might not wish to consult a junior wife etc. She did not mention men not wishing to go to a woman doctor but I wonder if this might also be a problem.[12]

Both examples reveal the unwritten system of beliefs that underpinned the FCO's social and cultural structures. The young doctor, mentioned above, was discouraged from applying for the post of medical adviser in New Delhi not because it was prohibited by any documented policy, but because it contravened tacit Foreign Office convention. After stating that fellow wives made her feel apologetic, Ibbott drew attention to this rigid yet informal structure which was based on an understanding of moral obligation. She ascribed other wives' opinions to 'attitude', noting that 'attitudes can curtail freedom of action as much as legislation, and indeed it is attitudes that working wives are up against … not legislation'. Ibbott's assessment of the power of 'attitude' to bring about conformity recalls Coles' remarks about the 'loyalty that began with the mission, but which implicitly tapped into a deeper, morally-invincible, loyalty to the Country and the Crown'.[13] Ibbott concluded that many diplomatic wives preferred not to seek work outside the mission and assumed unofficial and unpaid duties as a means of self-justification: 'The blunt truth is wives owe the post and office nothing at all.'[14]

Ibbot's letter was robustly countered in the succeeding newsletter by Jean Reddaway, a diplomatic wife and artist who often provided artwork for the DSWA magazine covers. Reddaway was keen to assert that she agreed with much of what Ibbott had to say but she moved the argument more firmly into family territory, interpreting Ibbott's statement that wives owed the Foreign

Office nothing at all to include families, Reddaway wrote: 'Can you really mean that, Margaret? I find it amazing and very sad if it is so. The Office has looked after our every need, with good housing, medicine, travel and domestic help.'[15] Reddaway's apparent expectation and acceptance of what Miller described as 'hierarchy, with its elements of paternalism and even authoritarianism' might point to a generation gap between her and Ibbott.[16] But what is especially interesting in terms of this study is that Reddaway felt that it was not only she and her husband who were intimately connected to the FCO, but her children. She felt that their involvement was 'after all a family affair' and she was happy to speak for them when she went on to describe the relationship she felt that her children had with the Foreign Office:

> Our children do not appear to have suffered. Perhaps a proof of this is that two of them have chosen to follow the same career, one as the wife of an FCO man and the other as a new recruit to the Office. The other cannot wait to leave school and try her luck in some form of overseas service.[17]

Ibbott's letter was greatly influenced by Hilary Callan, another diplomatic wife who was also a social anthropologist and author of 'The Premiss of Dedication: Notes Towards an Ethnography of Diplomats' Wives'.[18] This short, critical piece was written in 1975, when Callan was at post with her husband in a middle eastern city. 'The Premiss of Dedication' examined the official status of diplomatic wives and the nature of the authority that was exerted over them by the Foreign Office. Callan wrote that:

> Every [FCO] officer is subject to an annual report by his superiors. This report includes provision for comment on his wife's performance. A belief exists (denied by others but in my opinion quite well-founded) that their non-conformity to what is expected can lead to negative sanctions ...[19]

'The Premiss of Dedication' also identified the intangible and implicit nature of FCO demands and anticipated reactions to them. It suggested that wives who readily conformed to the demands of Foreign Office convention were not so much dismayed by Ibbott's challenge to the status quo as by the violation of an unspoken code of agreement and collusion, as though an acceptance of such a code is the first step in pledging allegiance to the Foreign Office. As Callan put it:

> she [the diplomatic wife] has accidental but no contractual links. To say this, however, is to put the matter very brutally, and in a way which (although it follows from the data) would be disowned with indignation by many of the very individuals who operate the system.[20]

Callan's analysis of the process that encouraged diplomatic wives to conform to the wider group was again formalised in the early 1980s in a volume of papers that she co-edited with fellow anthropologist Shirley Ardener. The concept of the 'incorporated wife' was developed by a group of feminist academics and defined by them as 'the condition of *wifehood* in a range of settings where the social character ascribed to a woman is an intimate function of her husband's occupational identity and culture'.[21] Male professions covered in the book that were understood to 'incorporate' wives included universities, the police force and the military, expatriate wives and those married to colonial administrators. Just over half the careers examined in individual chapters were based overseas, or involved significant amounts of international movement. It is clear that Reddaway embraced the supporting role of a diplomatic wife and appeared to have willingly been 'incorporated' by the system but crucial to this analysis is her confidence in writing on behalf of her children to whom she attributed a great enthusiasm for and sense of belonging to the FCO. The question of whether diplomats' children were in fact 'incorporated' into the Foreign Office has been key to this investigation.

Reddaway's enthusiasm when recounting her children's attachment to the Foreign Office and the life it engendered was a common habit among diplomatic wives in this period. In other evidence wives were seen to offer opinions on their children's experiences without even being asked. 'Have they suffered? On the contrary'; 'Our children do not appear to have suffered'; and 'My own three children would all say that they loved it'.[22] Their motives are complicated. Most simply, it is a feature of the human condition that mothers often feel entitled to speak on behalf of their children. But diplomatic wives, as mothers, appeared to use these kind of statements to reassure themselves that their children did not suffer from separation or other travails of diplomatic life and to emphasise how well they knew them. They also seemed to believe it was essential to report their children's spontaneously overt enthusiasm for and keen sense of belonging to the FCO system. However, contributors to this book who experienced the FCO as children significantly challenged this assumption. The way that they related to the organisation differed greatly from the relationship that their parents, including their mothers, had with it. Despite having lived in the shadow of the FCO from birth they were largely immune to its influence, rather than totally 'incorporated' as might reasonably be expected. Indeed, although diplomats' children were sometimes called on to represent their father's embassy in a small way, as we saw in Chapter One, they did not feel in any way 'incorporated' into the Diplomatic Service in the way that some wives have been said to be incorporated into their husbands' professions.

Children's voices

When the Foreign Service Wives' Association newsletter first appeared in 1961 it quickly became a space for wives to regularly air their preoccupations with boarding schools, holidays and the journeys in between, but the actual voices of their children were rarely heard. The first time the authentic voice of a Diplomatic Service child was heard in the newsletter was in 1963, in the form of a letter from ten-year-old Caroline Davidson – whose address was given as 'Avenue de Bourguiba, La Tunisie'. Caroline was writing to the FSWA to thank them for the parcel of books she had received from them. Her letter was similar in tone to that of a thank-you letter written to a distant family member after Christmas, and possibly benefitted from some adult prompting: 'I like mysterie and adventure stories very much, horsy books are nice too and I like biographies. I think your choice of books we would like is very good.'[23] Caroline also brings in members of her wider 'family', Tunisia's British community, her friend Roberta Weaver 'who hasn't got her books yet (they're on the way)' and 'the Knowltens (friends of ours)'.

The genuine voice of the Foreign Office child was rarely heard. The first section of this chapter made it clear that wives as mothers felt fully justified in speaking for their children, like Jean Reddaway in her letter to Margaret Ibbott, who confidently felt able to write 'Our children do not appear to have suffered'.[24] A similar assertion came from Anne Rothnie who declared, within the pages of the DSWA newsletter, that 'My own three children would all say that they loved it'.[25] Rothnie's enthusiasms, however, were printed as part of a sequence in the newsletter's Spring 1980 edition entitled 'Views of Children and Parents'. This item had been suggested in the preceding issue and was the first time that such a balanced picture had been presented. The reason for the sudden opening-up of the newsletter to FCO children is unclear; the explanation given in the magazine was that the feature was designed to coincide with the UN International Year of the Child, announced in October 1979. But the Foreign Office had always had an uneasy relationship with the kind of humanitarian internationalism represented by the UN. British diplomats to the UN and those based in London in the UN Department greeted the announcement of the Year of the Child celebrations with a mixture of apathy and cynicism. 'International years are expensive … international years have usually been ineffective … the whole exercise does seem rather pointless …' was the note on one exchange. 'After consulting home departments and noting views expressed by other members of the EEC, the UK sent a discouraging reply …' It was also feared that 'the Russians intend to exploit this item in the disarmament context'. The wording of article six of the UN Declaration on the Rights of the Child – 'A child of tender years shall not, save in exceptional circumstances, be separated from his mother' – caused some comment, mainly because FCO legal advisers felt that the UK might not be fully able to comply with it. A handwritten note explained the concern: 'because children of women in prison are taken away from their mothers at a point usually I believe between the ages of 2 and 3, where it is felt that the prison environment might be of harm.'[26] The enduring condition of family separation as a defining feature of Diplomatic Service life and its resounding irony in this context was not apparent to the note-writer.

The voices of children might have been introduced into the DSWA magazine in response to the growing interest in individual rights. Commenting on experience of childhood in the twentieth century, Thomson described 'The move, particularly in the last decades of the century, to see children not just as deserving subjects of care but also as having rights and a voice of their own …'[27] However, the sudden inclusion of FCO children's voices in the wives' magazine could be attributed to far less noble or complex reasons: for example a scarcity of contributions from the wives themselves and the desire of some parents to showcase their children's writing talents. The reader cannot ever be sure of the extent to which written evidence that has apparently been produced by children has been subject to adult intervention, or vigorous over-encouragement.

Perhaps the most curious thing about Anne Rothnie's quoted claim for her children's total enjoyment of diplomatic life came when she went on to write that they were all now over eighteen and the reader understands that she made the decision to answer a perceived question on their behalf: the magazine asked for contributions from children, not valuations of their life as the children of diplomats. Throughout the period of their existence, DSWA newsletters and magazines maintained a sharp focus on children's issues and gave the impression that FCO wives/mothers knew their children intimately, understanding how they felt and what they wanted. Diplomatic Service mothers felt that it was imperative to maintain a favourable account of their children's experiences, particularly in the case of separation. The wives, unsure or unhappy about the decisions they had made, might have been seeking public (although internal) reassurance that the diplomatic lifestyle did not harm their children, possibly because they suspected that it did. They might also have wanted to be seen to conform to the wider group in the way mentioned by Callan in her analysis of the 'incorporated wife', and felt more comfortable when expressing views that adhered to internal community norms. Their over-enthusiastic tone may have been assumed to mask a sense of guilt. Whatever the reason, it was very difficult for them: diplomatic wives, anxious about their children, concerned for their standing within traditional FCO culture and conscious of their loyalty towards their husband and his role within the organisation, were often in an impossible position.

These difficulties are echoed by the Foreign Office children who contributed to 'Views of Children and Parents' in the Spring 1980 magazine. What they wrote – or what is printed – indicates that the children were at great pains to place their experiences of Diplomatic Service life in a positive light, although the exception to this rule is the youngest contributor, David Blishen, a seven-year-old living in Tokyo, who was reported as saying 'Well I think it is not necessary for Mum to go out all the time! And I don't like babysitters. There shouldn't be so many parties in a year or even a week.'[28] The older children, however, whose commentaries focus on the positive, emphasise the same set of characteristics: they mention that being the child of a diplomat means that you see the world, you experience other cultures, you develop a lifelong interest in travel, 'the advantages are quite plain, discovery, knowledge, widening of scope and ideas and the satisfaction it brings'.[29] Because their common experience is one of boarding school, diplomats' children introduce the boarding school environment as though it is normative, setting great store on independence and being able to live with and get on with other people. 'One has lots of friends and learns how to adapt oneself to a large community', wrote one child, in a statement that could equally apply to the experience of working for the Diplomatic Service.[30] 'Being a boarder taught me a lot', says another. 'One is placed among strangers and left to fend for oneself. Tough as it is I want to stand on my own two feet and I learned a very important lesson in life – how to live with other people.'[31] Another contributor, Retta Bowen, questioned what she called a 'premium on independence', asking 'there's so much of a premium put on independence. Why is independence so good?'[32]

The children's writing indicated that they, like their mothers, felt it was necessary to excuse the situation in which they found themselves: they made the best of things and they often appeared defensive. Because of this, the children's contributions to the Spring 1980 newsletter are deeply ambivalent. Nine-year-old Rosalind Miller began cheerfully with 'I like being away from my parents because it is fun going to and fro from Belgrade to England'. Her tone soon changes, however, in the second and third sentences: 'I like going to friends at the weekend because it is nice to get away from school. Sometimes I feel homesick but I am always comforted by the other girls and they make me happy again.'[33] The contribution of M.N. Clarke, aged fourteen, is also deeply contradictory: 'I am at this present moment experiencing boarding school and my family live abroad therefore cutting me off from them for eight months of the year.' Clarke's use of the formal personal pronoun one, coupled with his relentless praise of boarding school and independence give a strong impression of desperate conformity and, ultimately, self-deception. For example the passage:

> Naturally one would be separated from one's family and at first this may seem a bad thing but I think it is on the whole a good one. One is much more independent and one is more capable of looking after oneself. For instance one often finds a person who has never travelled on a train before the age of fifteen, which is ridiculous and does not know how to make a decision without consulting his guardians. One learns the thing the hard way but that is better than being spoon fed like a hutch rabbit.[34]

Clarke concluded that he preferred boarding school to being with his family because he and his boarding school friends had more in common.

Like Clarke, Katherine Miller (the sister of Rosalind quoted above) used the pronoun 'one', which seemed excessively formal and might indicate the intervention of boarding school authorities keen on correct English in the production of these written pieces for the Foreign Office. Katherine Miller's long item is interesting for the instructions it contained, aimed at parents. For example, 'For those with daughters approaching puberty the advice is to tell them if you want (they probably already know anyway) and give them the necessary equipment'. Again in common with Clarke, Katherine Miller, indicated that she learned to rely on boarding school friends in her parents absence, stating 'They are perfectly able to fend for themselves and they have their friends'.[35] She also demonstrated a concern for parents, advising them not to worry about children flying alone, 'the constant chaperoning by air hostesses becomes boring', and revealing that she and her sisters 'never upset our parents while crying in their presence, we waited until we were out of sight'.[36] This is a significant example of both the sense of responsibility that Diplomatic Service life demanded from children and the way in which they developed methods of caring for their parents that reciprocated their parents' methods of caring for them. Children assuming protective roles towards adults was not uncommon. Anne Foster★ recalled a time when she was at post with her husband in New Delhi and was aware that her daughter was being bullied at her English boarding school: 'children don't want to worry you … [my daughter] said I once was so upset when she was upset on the phone that she vowed never to upset me again.'[37] Katherine and Rosalind Miller's mother Sheila provides an addendum to their written pieces, in which she states that 'All of our children have had behavioural problems (although not serious) which improved tremendously when we were at home on leave or posting'.[38] The improvement in children's well-being when their parents were closer to them calls to mind article six of the UN Declaration; these children were all separated from their mothers, although it might be argued that they are not 'of tender years'.

More than one family provided the views of several of its members. The female Millers, quoted above, were one, but most interesting in this respect are the Donalds, whose sons were aged between ten and eighteen. Alexander, who was the youngest, was also the most honest: 'I don't like being in the Foreign Office because Mum is never in England apart from the summer.'[39] His brother Angus, no doubt due to wider and longer experience, took a balanced and stoical view of his situation:

> One good thing for children whose parents are in the Foreign Service is that after a while, at each change of home, they begin to 'acclimatise' and they very quickly regard the new place as home. The variety of background of new homes not only prevents boredom, but when they start life as adults, they will have experience of other parts of the world.[40]

The oldest brother, eighteen-year-old John, was the most reflective. His written views begin in very much the same vein as those of M.N. Clarke who was quoted above (and there was a comparable use of the formal 'one' pronoun). Like Clarke, John Donald wrote that he felt that separation, transience and boarding school were necessary in the creation of resilience, fortitude and independence.

> It all comes down to the essential questions of pain and experience. Each successive uprooting, though painful at the time, makes one stronger, more independent and that bit more conscious of harsh realities that the sheltered child in the stable next is ignorant of.[41]

He also commented unconsciously on the absence of rebellion about Diplomatic Service children. 'The constant uprooting and the fact that being with parents is a rare and limited treat makes the teenage revolution much smoother when the child, inevitably, must cast off his parents and seek his own way …'[42]

John Donald's comments on the subject of friendship correlated with evidence gathered elsewhere. Donald noted that the Foreign Office child can:

> deal with both the moments of deep self-revelations to a close friend and the shallowness of too much social intercourse and the need for a social mask – these are natural parts of his upbringing. However between these two poles is the chink in his armour, coping with mid-distance relationships, 'relaxed, informal company' can be traumatic and awkward.[43]

A newspaper article credited with the dissemination of the term 'Global Nomad' listed among the limitations of the designation 'the trauma of constantly leaving behind people and places and a common distrust of emotional intimacy and long-term relationships …'[44] Many of the former Diplomatic Service children who contributed to this book reported difficulties with making and sustaining friendships, for the obvious reason that friendships were often broken when they moved on. One correspondent thought that this was a concern and a problem for children from a very young age:

> The worst thing was leaving my friends when we moved, particularly when I was about seven. That was when we went to live in Hong Kong and I left two very close friends in London and you kind of know that you won't really keep in touch …[45]

In the Spring 1980 DSWA magazine another older child, Katherine Ferguson, observed that the 'price' of being a Diplomatic Service child was 'security and stability of character' and that 'being the child of a diplomat can make or break you – at least the psychological effects can be serious'. But while naming all the advantages already mentioned, Ferguson adds another and that is that she enjoyed being a child of the Diplomatic Service because she liked the status.[46] This is something, which, although widely enjoyed, was not always so overtly stated. The final chapter, however, will testify to changes in the way diplomats perceived themselves and their status.

Diplomatic families and global threat

The late 1970s and early 1980s saw a rise in high-profile terrorist events in the West. It has been suggested that 'the entire framework of British foreign policy was determined after 1970 by the so-called "cold war"' and historians have been alert to links between the two, for example Blumenthau, who wrote: 'With a largely bipolar division of the international system, most terrorists in one way or another accommodated themselves to the Cold War.'[47] Rapid globalisation was another facilitating component in the advance of terrorism. The worldwide increase in foreign travel, especially by air, was not solely recreational: regional conflicts could now be played out on the world stage. A compelling example of this was provided by the Palestinian attacks on Israeli athletes at the 1972 Munich Olympics, or the 1980 Iranian Embassy siege in London. An additional factor in both of these incidents was that they received live television coverage: improvements in media communications meant that impact was instantaneous.

With this increase in high-profile violence for political ends, the later twentieth century saw a greater threat of brutality towards the Diplomatic Service family than anything previously recorded in peacetime. Repercussions of the troubles in Northern Ireland meant that diplomat Bill Cordiner, while serving as a Consul in the US, was regularly telephoned at home by representatives of the Irish Republican Army (IRA) who told his children when they answered the phone 'Tell your Dad we are going to get him'.[48] An inquiry into these incidences of violence and threat can tell us much about the role of the diplomat overseas during this time and exposes the level of resilience expected from Diplomatic Service families by the Foreign Office. It also gives an idea of what the British public might expect from Diplomatic Service families faced with these circumstances. While the first section of this chapter concluded that Diplomatic Service children did not feel pressured to 'incorporate' themselves into the Service the way their mothers did, two high-profile cases discussed in this section – the assassination of Christopher Ewart-Biggs in 1976 and the expulsion of British diplomatic families from Tripoli in 1984 – demonstrate the ways in which unforeseen episodes of threat forced Diplomatic Service children into representative or public roles.

Threats to individual diplomats, combined with the kinds of existing difficulties of diplomatic life discussed elsewhere in this book, had the power to construct dreadful situations for their children. Because families were separated, when individual diplomats came under threat overseas television news and other media coverage meant that their children heard the reports and feared for them greatly. The situation for parents, however, was more complicated and sometimes forced on them a different version of a choice with which diplomats were already familiar: that between family and duty. Peter Wakefield was posted to Beirut in 1975 during the Lebanese Civil War and was accompanied by his wife Felicity. Felicity later confessed to being 'hooked on danger' but her children suffered terrible anxiety:

> Our youngest son was still at boarding school. He was at Ampleforth and the monks said that his work was suffering. I think he was very anxious. You know, he heard things on the radio. I don't know what one ought to do as a parent. The children kept saying 'You ought to come home. You shouldn't be there.' I think they thought we should both come home.[49]

Older children were just as much affected. One contributor was in her early twenties when her father came under fire at a formal engagement in Cairo where Egyptian president Anwar Sadat was shot dead. Travelling on the London Underground she read the news as a newspaper headline over a fellow passenger's shoulder and burst into tears.

Children could be placed in dangerous situations without fully understanding their implications. Retta Bowen and her sister travelled with their parents to a posting in Lusaka during the late 1970s when the Rhodesian War of Independence was being fought along the Rhodesia–Zambia borders. Bowen recalled an example of the plain-speaking used by her family to depreciate the very real threat of sexual violence at the time: one which superficially called on the imperial tradition of moral fortitude and British sang-froid, the tradition that, according to Procida, inspired Anglo-Indian wives who 'willingly took on "unfeminine" roles as defenders of the empire'.[50] Bowen recalled:

> When my parents went out, we would be locked behind the rape gate. I was three, four and we would say 'Oh please don't lock the rape gate!' It's weird isn't it? ... We were aware in the house of there being curfews and violence and lorries going past with drunk soldiers on them ... I remember there being some sort of disturbance in the night and my Dad going out in his pants with a water pistol.[51]

Bowen's testimony indicated that despite the instability of the local situation her parents were required to go out – probably on diplomatic business – leaving their daughters behind. The nature of these recollections also poses the question why the Foreign Office felt it necessary to send families to posts like these at all.

This sense of authority, of taking events in one's stride, was successfully recalled by a former diplomatic child who was on a brownie camp in Seoul when the Korean president Park Chung-hee was assassinated in October 1979. Her account of returning to the British Embassy compound with a US army escort and watching her mother hang up wet towels as tear gas blew through their house was given in a very matter of fact way, without drama and with some humour.[52]

The way in which the public and private spheres were blended for diplomats serving overseas could be confusing. Mori observed that 'Diplomacy has always been a lifestyle requiring its disciples to abandon customary distinctions between public and private life ...'[53] The recollection of tear gas blowing through the house in Seoul demonstrates that the home life of diplomats overseas continued to diverge from common middle and upper class practice in Britain well into the twentieth century. The 'advancing home-centredness' described in earlier chapters as characterising post-war Britain, accompanied by the respectable notions of 'keeping yourself to yourself', was not possible for diplomatic families serving overseas during the same period, even if they had been the inclination of the individual family.[54] Current Diplomatic Service Regulations make clear to the concept of 'all hours liability':

All Diplomatic Service staff have an all hours obligation to the FCO while they work overseas. That means that they must be prepared to make themselves available to their Post management at any time of the day or night seven days a week.[55]

In July 1976 the new home of Christopher Ewart-Biggs' family became the site of violence and disaster. Ewart-Biggs was killed by an IRA bomb two weeks after arriving in the Republic of Ireland to take up the post of British ambassador. His three children were home from school for the summer holidays, but his wife Jane was in England to buy decorating materials for their new home. The Ewart-Biggs' youngest child, Kate, remembered spending time with her father that morning before he left for work. When he had gone:

> I wandered outside to find Owen, my new friend … and we set about a game in the garden. We were very soon stopped in our tracks as we heard a large bang and felt the ground shake – we looked at each other for a minute, shrugged our shoulders and carried on playing – I learned later that the noise was a mine being detonated underneath my father's car. It killed him … instantly.[56]

The family of Katie Hickman, who wrote the popular book about diplomatic wives, *Daughters of Britannia,* was also in Ireland because her father John Hickman was deputy head of the British mission, thus Ewart-Biggs' second in command. Hickman's wife Jennifer wrote a compelling diary account of the days surrounding Ewart-Biggs' death, excerpts of which appeared in her daughter's book. Notable is the way that Jennifer Hickman, who had not known the Ewart-Biggs family for very long, immediately assumed responsibility for them and accepted the task of telling the two older children that their father was dead. The Foreign Office view of itself as a wider family was reinforced: but also stressed was the challenging nature of the quasi-official duty given to Jennifer Hickman. The description of the immediate aftermath of Ewart-Biggs' death also raised powerful questions surrounding the transient aspects of Diplomatic Service lives. Contributors to this book named transience (which was defined as a perceived or actual state of constant movement and upheaval) as the second most challenging aspect of diplomatic life after separation. An analysis of a random sample of ten contributing families showed that, on average, in a thirty-five-year diplomatic career, families were posted overseas eight times. The children of those few families who used international schools described their experience as being like a revolving door and the structure of Diplomatic Service recruitment meant that families' lives could undergo profound and immediate change every eighteen months to two years. The experiences of the Ewart-Biggs family are, of course, exceptional, but the brutal fact of their father's murder so soon after the family's arrival in Ireland gives a further, shocking dimension to the unsettling (and often bewildering) nature of a transient life. Diplomatic children felt very strongly that the demands of this transience exhausted their mothers' resourcefulness, illustrated by the burdens placed on Jennifer Hickman that day.

In extreme situations such as those surrounding Ewart-Biggs' death, the Diplomatic Service family home experienced an invasion of privacy that was made worse because the boundaries between their public and private lives were obscured. Jennifer Hickman's diary account noted 'Little Katie asks innumerable questions as to why there are so many soldiers, policemen, armoured cars, helicopters, cars, people, etc, everywhere'.[57] Kate Ewart-Biggs was not told of her father's death at the same time as her older siblings but was curious about the reasons why her home had been invaded by so many different people:

> I remember knowing that there was something going on and being cross that they wouldn't tell me. I came to the conclusion that my mother had been killed in an aeroplane accident and, after a long day of pitied looks and hushed tones, took myself outside to ride my bike.[58]

Soon after their father's death, the Ewart-Biggs children appeared on a televised broadcast from the ambassador's residence. Kate Ewart-Biggs recalled:

> my sister Henrietta, 15 and desperately self-conscious, wanting to be anywhere but standing in front of the camera; my brother Robin, who, having turned 13 five days after losing his father, had to look like he was now taking on the responsibility of this family; and me desperately striving to look positive with a little shy smile.[59]

Contemporary news coverage showed that the children also gained the sympathy of the crowds gathered at memorial services for their father in London and Dublin. These appearances on the national stage at formal, ceremonial occasions, and the emotional resilience and sophistication that they call for, are experiences that only very few children – with the exception of children from the British royal family – will ever have.

Another cogent illustration of the way international events could bleed into private and family life of diplomats was provided by accounts of the evacuation of Diplomatic Service families from Tripoli following a breakdown in diplomatic relations between the UK and Libya in 1984. When in April that year a young British police officer, Yvonne Fletcher, was shot dead while overseeing a demonstration outside the Libyan People's Bureau in London, British diplomats in Tripoli were initially put under house arrest before they were given a week to leave the country. During the highly volatile and confusing period leading up to the British departures, 'cars containing revolutionary guards were placed outside ... six houses in the suburbs of Tripoli occupied by Embassy staff', telephone lines to the ambassador's residence were cut and staff were prevented from leaving the Embassy building.[60] Julia Miles, wife of the British Ambassador, was trying to sell and give away as much as she could from the residence (she had heard that mobs often ransacked official buildings previously occupied by western powers) but these developments gave her pause for thought. 'I did not feel like driving all over Tripoli ...' she recalled, 'I was apprehensive about leaving the children at home, just in case I was held and we were separated.'[61] As in the Ewart-Biggs case described above, the children of the British families were at home for the Easter holidays:

> They were very frightened. Even our eldest son who ... became a royal Marine and is tough ... he actually came in to the bedroom at night – I mean Oliver was locked up in the embassy and I was locked up in the house ... – in the very small hours of the morning and said 'Is it going to be all right?' and I said 'It's going to be fine.' But of course none of us knew how it was going to turn out at all we were just living day by day ...[62]

An account of the events in Tripoli written by a diplomatic wife, Barbara Cooling, which was printed in the DSWA magazine, brought attention to the highly gendered aspects of diplomatic life, even in or perhaps especially in times of crisis. Cooling described how, when the British Embassy in Tripoli was surrounded by Libyan security police and its mainly male occupants were prevented from leaving, 'they had been very well looked after by three wives who lived in the flats above the building... [the wives] cooked enormous meals, laid on baths, arranged beds and generally provided a haven for those who were detained'.[63] It is quite clear from this passage that the provision of home comforts was understood to be a feminine duty, the preserve of wives. Women and children were eventually instructed to gather at the residence in the hope of leaving for UK together and Cooling noted: 'None of us really wanted to leave our husbands, but we knew it removed a large burden from the men, who had more than enough to do at the Embassy, if we were all safely out of the way.'[64] While the wives' role was to care for the menfolk and children, men's work could only be seriously dealt with if the 'burden' of their families was removed. As with Retta Bowen's experience of Lusaka during the civil war, the case of the Tripoli evacuation raises questions about the suitability of some posts for families. The existing diplomatic methods of doing business, through what Simon Jenkins referred to as the 'Prism of bourgeois entertaining', was in many cases no longer appropriate or useful. Nonetheless, the wives who left their husbands behind were faced with burdens of their own: as they touched down in London, wrote Cooling, they experienced some 'apprehension about the new problems we had to face on our arrival ... we had to turn to the practical problems of transport, accommodation, schooling for our children, there certainly was no time to relax.'[65]

Julia Miles was intensely critical of the Foreign Office response to the evacuation of British families from Libya. The women, with their children, had had a long wait at the airport in Tripoli, so that the Libyan authorities could make sure that their diplomats had been permitted to leave London. Miles angrily recalled the moment when the women finally arrived at Heathrow, to be greeted by a bank of press photographers and journalists:

the Foreign Office were just shocking. I said … nobody rang up from the Foreign Office to ask how I was throughout the whole time and [they] said 'We didn't know we could dial direct.' That was [the] response. You know … real care and concern I have to say, I thought it was absolutely unacceptable behaviour.

They had a sort of reception laid out and all these lovely ladies from Welfare Department in black lace sort of saying 'Are you all right, are you all right?' … what I wanted was someone to say 'I'll hold the baby.' And then they said 'Here's your mail.' And they gave me it all loose and I had two children and all this … I mean, all I needed was a plastic bag … and they were saying 'Would you like a gin?' You know – fuck off! I don't want a gin.[66]

Miles' reflection on the evacuation of British families from Tripoli in 1984 led her to make a perceptive observation about Foreign Office culture that can be linked back to Margaret Ibbott's remarks in the DSWA magazine almost a decade before – discussed in the first section – about the intangible force that she described as an expression of 'attitude' and which acted as an effective restraint to working wives. Attempting to sum up the way in which the eva-cuation of her family and others was handled by FCO authorities, Miles said 'The Foreign Office is not a practical organisation, you know, it's an intellec-tual organisation … and when they're faced with something human they fall to pieces, they don't know what to do …'[67] The diplomatic wives' recollections of the evacuation from Tripoli revealed that the Foreign Office response to the human element of this crisis was inadequate – it involved some very small children and Miles' youngest daughter was suffering from chicken pox. This raises the question why so much emphasis was placed on family within the organisation, why families were permitted to travel to posts which could become dangerous and why there was no contingency for this possibility within the FCO's structure. There is little evidence to support any substantial conclusion but the failure to recognise that some posts had become too dan-gerous to house families is another example of the lack of interest in family and an outmoded outlook.

Conclusion

The period 1972–1985 saw changes in the expectations of FCO wives. This had important links in three areas. The first was that, as graduates of new uni-versities married their comparably qualified peers, educated women, some with professional qualifications and experience, were not prepared to take a back seat for their husbands. Second, the number of women entering the UK work force was on the rise and this was reflected in the aspirations of Diplomatic Service wives. Third, the marriage bar which had meant that female members of For-eign Office staff were required to resign on marriage was lifted. This meant that for the first time women could be diplomats, wives and mothers. However, diplomatic expatriate life continued to be highly gendered into the 1980s as the accounts of British families' expulsion from Libya following the breakdown of diplomatic relations in 1984 illustrates.

The second section of this chapter discussed the introduction of children's voices into the DSWA magazine for the first time in the spring of 1980, ostensibly as part of the UN International Year of the Child. Evidence showed that the Foreign Office did not view plans for the International Year of the Child in a positive light. It discouraged British diplomats at the UN from showing enthusiasm, giving the reasons that such a celebration would be expensive and that it could potentially be used by Cold War antagonists to draw attention away from what the British considered to be more significant issues. Foreign Office officials, well versed in the Foreign Office culture of family separation, who commented on a draft of the UN Declaration of the Rights of the Child, failed to see the irony inherent in article six which recommended that children 'of tender years' were not separated from their mothers. This points to an important dichotomy between the experiences and expectations of Foreign Office children and children from other groups. The views of Diplomatic Service children that were printed in the DSWA magazine mimicked those of their mothers elsewhere in their relentless expression of positivity. All children included identified similar advantages of being a diplomatic child and more complex issues – such as children's concern for parents during separation and the difficulties of forming lasting friendships – were corroborated by evidence gathered elsewhere.

The period covered in this chapter shows that diplomats and their families were put at greater risk than ever before owing to an increase in global violence, often for political ends. Families were placed in positions of very serious risk and it must be noted that the way the Diplomatic Service was structured (e.g. husbands accompanied by their wives and young families) did nothing to reduce this. The families and the risks that they encountered who have been described in this chapter would not have experienced such a high level of threat had the husband/ father not been a diplomat. A specific consequence of threat and violence on the family was the appearance of the three bereaved Ewart-Biggs children on national television, during a televised broadcast from their own home shortly after their father's assassination and when they attended memorial services for him in London and Dublin. This level of publicity and its accompanying demand for the assumption of a public 'face' and an expression of highly developed resilience is something that is only usually seen as experienced by children from the British royal family. Julia Miles' account of her return to London from Tripoli in 1984 demonstrates that existing FCO welfare and pastoral systems were not sufficiently developed to cope with the consequences of these incidents.

Notes

1 'What Is In It For Me?', *DSWA Magazine*, Autumn 1985 Commemorative Edition.
2 Simon Jenkins and Anne Sloman, *With Respect Ambassador. An Inquiry into the Foreign Office* (NP: BBC, 1985).

3 'With Respect, Ambassador, a Conversation with Simon Jenkins', *DSWA News-letter*, Autumn 1984.

4 Ibid.

5 See Harry Hendrick, *Children, Childhood and English Society 1880–1990* (Cambridge University Press, 1997), 97, for a short discussion on this.

6 Helen McCarthy, *Women of the World: The Rise of the Female Diplomat* (London: Bloomsbury, 2014), 286.

7 Helen McCarthy, 'Women, Marriage and Work in the British Diplomatic Service', *Women's History Review* 23:6 (2014): 853–873.

8 Sheila Skinner in Alun Parry, *Wives Do Not Work – A History of the Marriage Bar* (PCS Union, 2012). Film.

9 Anne Foster★. Interview by Author. 11 February 2015.

10 Margaret Ibbott, 'The Role of the Foreign Service Wife: A Personal View', *DSWA Newsletter*, Spring 1976, 43–45.

11 Ibid.

12 Smedley Papers, Box B.

13 Anne Coles, 'Making Multiple Migrations. The Life of British Diplomatic Families Overseas', in *Gender and Family among Transnational Professionals*, ed Anne Coles and Anne-Meike Fechter (London: Routledge, 2008), 125–148.

14 Ibbott, 'The Role of the Foreign Service Wife', 45

15 Jean Reddaway, 'Open Letter to Margaret Ibbott', *DSWA Newsletter*, Autumn 1976, 52–53.

16 Eric Miller, 'Some Reflections on the Role of the Diplomatic Wife', in *From Dependency to Autonomy: Studies in Organisation and Change* (London: Free Association Books, 1993), 136, first published in the DSWA Newsletter, Spring 1978.

17 Reddaway, 'Open letter to Margaret Ibbott'.

18 Hilary Callan, 'The Premiss of Dedication: Notes Towards an Ethnography of Diplomatic Wives', in *Perceiving Women*, ed. Shirley Ardener (London: Malaby Press, 1975), 87–104.

19 Callan, 'The Premiss of Dedication', 103.

20 Ibid., 97.

21 Hilary Callan and Shirley Ardener, *The Incorporated Wife* (London: Croom Helm, 1984), 1.

22 Frances Wooller, Questionnaire completed for *Partners in Diplomacy*, undated, late 1980s, Smedley Box A; Reddaway, 'Open Letter to Margaret Ibbott', 52–53; Anne Rothnie, 'View of Children and Parents', *DSWA Newsletter*, Spring 1980, 27.

23 Caroline Davidson, 'Letter to the Editor', *FSWA Newsletter*, 7 July 1963.

24 Reddaway, 'Open Letter to Margaret Ibbott', 52–53.

25 'Views of Children and Parents', *DSWA Newsletter*, Spring 1980, 27–31.

26 All comments from correspondence in TNA FCO 58/1490 International Year of the Child (IYC): Proposal for a Convention of the Rights of the Child 1979 Jan 01–1979 Dec 31.

27 Mathew Thomson, *Lost Freedom: The Landscape of the Child and the Postwar Settlement* (Oxford: Oxford University Press, 2013), 4.

28 David Blishen, 'Views of Children and Parents', *DSWA Newsletter*, Spring 1980, 32.

29 Catherine Ferguson, 'Views of Children and Parents', *DSWA Newsletter*, Spring 1980, 32.

30 MN Clarke, 'Views of Children and Parents', *DSWA Newsletter*, Spring 1980, 28.

31 Katherine Miller, 'Views of Children and Parents', *DSWA Newsletter*, Spring 1980, 30.

32 Interview with Retta Bowen. 22 May 2015.

33 Rosalind Miller, 'Views of Parents and Children', *DSWA Newsletter*, Spring 1980, 28.

34 Clarke, 'Views of Children and Parents', 28.

35 Katherine Miller, 'Views of Children and Parents', 29.

36 Ibid.

37 Anne Foster*. Interview. 11 February 2015.
38 Sheila Miller, 'Views of Children and Parents', *DSWA Newsletter*, Spring 1980, 30.
39 Alexander Donald, 'Views of Children and Parents', *DSWA Newsletter*, Spring 1980, 30.
40 Angus Donald, 'Views of Children and Parents', *DSWA Newsletter*, Spring 1980, 30.
41 John Donald, 'Views of Children and Parents', *DSWA Newsletter*, Spring 1980, 30.
42 Ibid.
43 Ibid.
44 Peter Kingston, 'Cosmopolitan but rootless', *The Guardian*, 18 May 1993. Kingston writes that the term 'Global Nomad' was invented by Norma McCaig: 'A Fellow global nomad … who founded Global Nomads International, a non-profit-making organisation based in Washington DC.' The term has a more romantic sound than Third Culture Kid but essentially describes the same group of young people whose parents' work takes the family around the world and causes frequent arrivals and departures.
45 Interview with Kate Howells. 22 July 2014.
46 Catherine Ferguson, 'Views of Children and Parents', *DSWA Newsletter*, Spring 1980, 31.
47 Bernhard Blumenau, 'The Other Battleground of the Cold War: The UN and the Struggle against International Terrorism in the 1970s', *Journal of Cold War Studies* 16:1 (2014): 61–84.
48 Bill Cordiner, *Diplomatic Wanderings From Saigon to the South Seas* (London: Radcliffe Press, 2003), 194.
49 Katie Hickman, *Daughters of Britannia: The Lives and Times of Diplomatic Wives* (London: Flamingo, 1999), 270.
50 Mary A. Procida, *Married to the Empire: Gender, Politics and Imperialism in India 1883–1947* (Manchester: Manchester University Press, 2002), 219.
51 Interview with Retta Bowen. 22 May 2015.
52 Questionnaire. Antonia Mochan. 8 August 2014.
53 Jennifer Mori, *The Culture of Diplomacy* (Manchester: Manchester University Press, 2010), 17.
54 Brian Harrison, *Seeking a Role: British Society 1951–1970* (Oxford: Clarendon Press, 2009), 190, 213.
55 Diplomatic Service Regulations (Internal FCO document), DSR 17, Hours of Attendance, Overseas.
56 Kate Ewart-Biggs, *Skiing Uphill*, British Council publication, February 2020.
57 Hickman, *Daughters of Britannia*, 256.
58 Ewart-Biggs, *Skiing Uphill*.
59 Ibid.
60 Barbara Cooling, *Evacuation from Tripoli*, DSWA Magazine, Autumn 1984, 59–64.
61 Julia Miles, *The Ambassador's Wife's Tale* (London: Eye Books, 2015).
62 Interview with Julia and Oliver Miles. 29 September 2014
63 Cooling, 'Evacuation from Tripoli', 59–64.
64 Ibid.
65 Ibid.
66 Interview with Julia and Oliver Miles 29 September 2014.
67 Ibid.

Bibliography

Beryl Smedley Archive. Archive Box B: Press cuttings and FSWA/DSWA material relating to diplomatic wives.

Blumenau, B. 'The Other Battleground of the Cold War: The UN and the Struggle against International Terrorism in the 1970s', *Journal of Cold War Studies* 16:1 (2014): 61–84.

Callan, H and Ardener, S. eds. *The Incorporated Wife* (London: Croom Helm, 1984).

Callan, H. 'The Premiss of Dedication: Notes towards an Ethnography of Diplomats' Wives', in *Perceiving Women*, ed. S. Ardener (London: Weidenfeld and Nicolson, 1975).

Coles, A. 'Making Multiple Migrations: The Life of British Diplomatic Families Overseas', in *Gender and Family Among Transnational Professionals*, ed. A. Coles and A. Fechter (London: Routledge, 2008).

Coles, A. and Fechter, A. eds. *Gender and Family Among Transnational Professionals* (London: Routledge, 2008).

Cordiner, B. *Diplomatic Wanderings, from Saigon to the South Seas* (London: Radcliffe Press, 2003).

DSWA Newsletter Spring 1976; Spring 1980; Autumn 1976; Autumn 1984; Autumn 1985.

Ewart-Biggs, K, *Skiing Uphill*, British Council publication, February2020.

FCO Files 58/1490 *The National Archives, Kew*.

FSWA Newsletter July 1963.

The Guardian, 'Cosmopolitan but rootless', 18 May1993.

Hendrick, H. *Children, Childhood and English Society 1880–1990* (Cambridge: Cambridge University Press, 1997).

Hickman, K. *Daughters of Britannia: The Lives and Times of Diplomatic Wives* (London: Flamingo, 1999).

Jenkins, S. and Sloman, A. *With Respect Ambassador: An Inquiry into the Foreign Office* (NP: BBC, 1985).

McCarthy, H. *Women of the World: The Rise of the Female Diplomat* (London: Bloomsbury, 2014).

McCarthy, H. 'Women, Marriage and Work in the British Diplomatic Service', *Women's History Review* 23:6 (2014): 853–873.

Miles, J. *The Ambassador's Wife's Tale* (London: Eye Books, 2015).

Miller, E. 'Some Reflections on the Role of the Diplomatic Wife', in *From Dependency to Autonomy: Studies in Organisation and Change* (London: Free Association Books, 1993).

Mori, J. *The Culture of Diplomacy* (Manchester: Manchester University Press, 2010).

Parry, A. *Wives do not work – A History of the Marriage Bar* (PCS Union, 2012). Film.

Procida, M.A. *Married to the Empire: Gender Politics and Imperialism in India 1883–1947* (Manchester: Manchester University Press, 2002).

Thomson, M. *Lost Freedom: The Landscape of the Child and the Postwar Settlement* (Oxford: Oxford University Press, 2013).

4   1985–1990

During the period covered in this, the final chapter, public sector workers in the UK were hit by a range of unpopular cost-cutting policies, referred to by the Foreign Office Chief Clerk as 'the realities of the 80s'. Margaret Thatcher's Conservative governments translated that Prime Minister's wariness of Whitehall civil servants into a campaign of cuts and reforms. 'During her first nine years as prime minister, civil service numbers fell by more than a fifth, and between 1979 and 1993 more than a million public sector jobs were transferred to the private sector.'[1] Foreign Office management put it in this way: 'At the heart of that change has been the commitment of successive conservative governments … to lower inflation, cut taxation and reduce government spending in the division of the national cake.'[2] But these exercises in austerity struck at the very heart of the Diplomatic Service, which feared a loss of identity as well as a decline in influence and status and in basic living standards. The DSWA Magazine of Spring 1986 bitterly noted that 'Most of our complaints against the Service could be summed up by saying "we're fed up with being treated as though we were members of the Home Civil Service"'.[3] In 1988 the Foreign Office Chief Clerk, Sir Mark Russell, made an ominous address on the changing nature of diplomatic society:

> Specifically I would like to talk more about the changing environment with which the Service, the Diplomatic Service Administration and you yourselves have to contend and which so much governs the way we tackle our specific problems, whether those problems are of office organisation, methods of work, conditions of service, establishments, our property at home and overseas, furnishing healthcare travel and so on. Because, make no mistake, the environment is changing rapidly.[4]

Russell went on to recognise that the 'policy of pay restraint' was 'giving rise to severe problems, particularly at the junior levels of the Service'.[5] He was, however, equally quick to remind members of the Diplomatic Service that they were still seen as part of an elite: 'whether we like it or not, we are regarded as a privileged part of the public service.'

There were other fears in the diplomatic world. Diplomats' wives, who had formed an effective 'shadow' hierarchy alongside their husbands throughout the

DOI: 10.4324/9780429273568-4

twentieth century, were being forced to reassess their roles inside the Foreign Office and outside it. Since the 1972 abolition of the marriage bar women diplomats had occupied those roles that had been reserved for men and the Diplomatic Services Wives Association (DSWA) had seen greater numbers of male spouses among its members, which would result, ultimately, in a change of name. The traditional, gendered system of values and behaviours cherished by the Diplomatic Service was under assault. Divorce had become acceptable and widespread: according to Clarke, during the 1980s there were 'over 160,000 decrees a year, not far short of half the number of marriages taking place'.[6] This meant that larger numbers of young people grew up in one parent households headed by women or in 'blended' families, where parents' remarriages introduced a step-parent and possibly step-siblings.

The previous chapter discussed the ways in which employment had become a concern for wives and it was during the period 1985–1990 that women, as part of the UK workforce, began to dominate new sectors, such as the emerging service industries. While their mothers were better employed, children in Britain remained dependent for longer than ever: '"Childhood" filled the long stretch from the start of schooling ... into teenage years ...'[7] But welfare cuts did not make young lives easy. In 1988 the majority of sixteen- and seventeen-year-olds lost their entitlement to social security benefits, while two years later students lost their right to housing benefit and the student loans system was introduced in lieu of the means tested maintenance grant.

But for all these changes in British society and in the Diplomatic Service itself, many problems affecting Foreign Office families remained similar, if not the same, to those that they encountered after the Second World War and change in some areas of diplomatic society was as slow and circumspect as ever. This chapter's first section assesses those aspects of diplomatic life that were slow to change – and which indeed had perhaps not changed since the post-war years covered in Chapter One – ways of life which presented a marked difference from British social practice at the end of the twentieth century. The second section will examine the fundamental change in attitudes (primarily exhibited in the DSWA magazines) towards family issues alongside substantive changes to social and welfare policy that finally began to loosen entrenched Diplomatic Service routines. The final section considers the changes in lifestyle and corresponding new identities sought out by western transnational professionals and their children as they made sense of their roles as western migrants in post-colonial settings.

Plus ça change, diplomatic family life in the late 1980s

In 1986 the DSWA magazine printed an advert for childcare services offered by Mrs Margaret Morris which read 'Escort service for children. To and from schools, railway stations and airports. Overnight service if needed'. There was a private address given: '8 Monksdene Gardens, Sutton, Surrey.'[8] This advert was typical of many that had been printed in the DSWA magazine since its inception in 1961. The bad reputation of holiday homes among families whose parents were engaged professionally overseas was touched on in Chapter One. Nonetheless evidence given to the Plowden Committee in 1962 by the FSWA suggested that one in four diplomatic families had made use of them because there had been no one else available to take care of their children during school holidays or transfers from post to school. There is no reason to suspect any ill of Mrs Morris' intentions; it is likely that she had run the service efficiently for years and probably had some connection with the Diplomatic Service or other public service, as many escort services and holiday home proprietors who advertised with the Diplomatic Service did. However, the appearance of her advertisement in the DSWA magazine in the same year that saw the launch of Childline, a UK charity established to help young people suffering abuse, and which quickly became '*the* place that children and young people identify as their own' serves as an indication of how far Diplomatic Service views of children, their care and implications for their sense of agency, had departed from those of the British mainstream.[9]

By 1986, national charities like Childline and private ventures like those of Mrs Morris occupied polarised ends of a spectrum regarding attitudes to children and their care. Childline had its origins in a BBC television programme that aired a survey about child abuse in the UK: when the helpline provided after the show was inundated with calls its life was indefinitely prolonged. Linked to the growing emphasis on children's rights and given impetus by the 1979 International Year of the Child, Childline and related charities brought into the open under-discussed and under-represented areas of children's experience with an aim to combatting cruelty and maltreatment. Childline was forward thinking, an idea that had never been put into practice before; it made use of the (limited by today's standards) technology available in 1986 to disseminate its message through television broadcasts and opened a free telephone helpline manned by qualified personnel which could be accessed via an number that was easy to recognise and remember.[10] Crucially, children and young people in difficulties were encouraged to contact Childline direct, giving the details of their own story in their own way.

In comparison, ventures like the one that offered Diplomatic Service families childcare with Mrs Margaret Morris were outdated in almost every sense. Margaret Morris was an individual – and presumably a stranger to the diplomatic children she was offering to care for in her own home. The advertisement contains no reference to testimonials or membership of professional childcare bodies. Its appearance in a Foreign Office in-house magazine suggests that the Diplomatic Service still recognised and relied on the informal systems of childcare that had prospered on the sidelines of colonial life. The ad hoc nature of the service offered by Margaret Morris is an indication that within the 'total' structure of the Foreign Office children were still viewed as a peripheral concern. Commenting on the practice as it grew up around 'Raj Orphans', Brendon writes:

There were certainly no checks into the suitability of these guardians, on whom many expatriate parents now relied. The dangers of private fostering, which is still common practice today, were emphasised in a 1997 government report. It concluded that unregulated fostering puts children at 'very considerable risk'...[11]

Without a doubt, the possibility of informal childcare structures would not have been an issue if family separation through boarding school had not still been the normative experience for the majority of British diplomats' children. As late as 1990, at the very end of the period covered in this book, the DSWA magazine – soon to change its name to encompass the number of diplomatic husbands now linked to the organisation – ran another article on 'Choosing Boarding Schools'.[12] The families of these children at school in the UK made use of friends from the Foreign and Commonwealth Office (FCO) network and other childcare schemes that were run by the DSWA. Children on their way back from holidays at post with their parents recalled sitting in unfamiliar kitchens as part of schemes run by the DSWA where they were collected and entertained by FCO families. More often, however, it was family and friends who helped children move from one place to another or kept them amused for a spare day. Contributors were often well aware that separation, especially in its boarding school manifestation, had placed them in a questionable situation:

> there were resident nurses in each house, and the woman that was in the one in our first house when I first went to boarding school used to be in a mental institute ... There were things that happened ... we all had to have checks because they didn't want people getting eating disorders and all that sort of thing and we had this weird check when we were probably only about eleven or twelve. We all had to go just in our knickers to see the doctor and looking back on that and even at the time, that felt slightly not right somehow that we were all lined up with our dressing gowns on and only pants on underneath to go and be investigated ... that wouldn't be happening if you were at home ... I think it was easier even for people whose parents were in the country. Even if you have guardians – I didn't really know my aunt and uncle well enough to say 'I think it's a bit weird that we go in just our pants to see the doctor'.[13]

This continued stress on family separation during the 1980s was at variance with national habits. Bowlby and Winnicott's teachings had long been absorbed into mainstream childcare orthodoxy and separation was out of the ordinary. The number of boarders at independent schools in the UK fell from 131,000 in 1975/6 to 94,000 in 1992/3, while public sector funding cuts led to closures of those boarding schools run by local education authorities.[14] Anne Foster*, both employed by the Diplomatic Service and married to a diplomat, recalled the 'absolute agony' she felt saying goodbye to her children, young in the 1980s, as they left for boarding school.

A lot of people were beginning to say 'How can possibly put your children in boarding school can you not, you know, have them with you?' And I think, probably, thinking back, now if I had my time again I might think a lot harder about it. Because I was married to someone who'd been to boarding school from the age of seven and so for him there wasn't a question about it.[15]

Foster★ found herself bound by the irresistible logic of tradition. She was concerned that the pressures of a transient lifestyle could be as damaging as separation and that international qualifications like the international baccalaureate – which were still relatively unusual – would not be sufficient to allow young people to enter higher education in the UK. Perhaps most significant is that Foster's husband at the time, himself a former diplomatic child, had experienced long periods of time away from his own family as a young person and had not expected anything different for his own children. The Foreign Office set great store by the British public school system as standard for its children. Public schools were referred to constantly in Foreign Office literature (the DSWA magazines for example) and during the status-conscious 1980s the Diplomatic Service allowance that met the cost of school fees begin to be seen as an essential part of the role, fulfilling expectations of diplomacy as an elite international lifestyle. The perception of 'the good life' was pervasive; recruits to the Diplomatic Service around this time remember being attracted by the promise of travel and embassy parties.

Steiner wrote that 'diplomats … have been forced to take part in self-examination exercises' and while this is true of the compulsory participation in structural reviews like that conducted by Plowden in the early 1960s they are not so applicable to their social and family lives. It is difficult to determine why the culture and image of the Foreign Office remained static for so long. Anne Coles' 'Making Multiple Migrations' is one of the few academic studies that goes some way to address this question, stating that 'there was a period in the 1970s and 1980s when the FCO and its members seemed to be about fifteen years out of date compared with the rest of the country'.[16] Coles conceded that the Foreign Office authorities were slow to respond to the need for reform specifically in relation to gender and family issues. She pointed out that by the time diplomats reached leadership level they were at least a generation older than their younger, more junior, colleagues and felt that the absence of coordinated reform was due to the 'far-flung nature of diplomatic posts' and the fact that, when overseas, Diplomatic Service families were isolated from advancements in mainstream British life.

Two of Coles' suggestions, above, are worth further attention. First, the generation gap within the FCO hierarchy. This is a very interesting point, but an important lesson can be learned from an analysis of the tone of the DSWA newsletters written in the late 1980s. These were generally written by younger wives or those at the 'mid-point' of a Foreign Office career and what is striking about them is how old-fashioned they sound, with contributors referring to their children public school fashion as – for example – 'Larmour One, Larmour Two'. It is likely that the Foreign Office attracted young recruits who were outmoded and conservative in attitude. The reliance on and partiality for other establishment institutions expressed in internal Foreign Office literature feeds into the overall sense of conservatism. Possibly, young recruits to the Diplomatic Service suffered from what Steiner described as 'a tendency for some Foreign Service sons to follow in their father's footsteps', coming from backgrounds with a tradition of or belief in duty and sacrifice.[17] Equally significant is Coles' point that diplomats and their families who spent a lot of time overseas were out of touch with contemporary UK life: the type of sentiment that Steiner describes as 'Britain's unfortunate love affair with its Victorian past'.[18] Contributors felt that the Diplomatic Service – especially overseas – presented an outdated vision of Britain that was in keeping with the behaviours and allegiances discussed above. The most popular diplomatic 'type' appeared to be the British 'eccentric' who gave the impression of having been at Eton, possibly because they fulfilled pre-conceived notions about Britishness and British behaviour: they became what they were expected to be. The popularity of the BBC television comedy *Yes Minister* in the 1980s created a similar situation; the writers of the programme were influenced by the behaviour of civil servants who, in turn, were influenced by the programme. Harrison observed: 'Civil servants … watched the programmes, feeding ideas into them and getting ideas from them.'[19]

The 'persistent gendering of expatriate lives' and gender roles within Diplomatic Service families remained fixed well into the period under discussion.[20] During the 1980s a space to comment about an officer's wife on his Annual Confidential Report (precursor of the Staff Appraisal) remained, a much discussed and controversial practice which the Administration maintained was to monitor the wife's health and well-being. In general, despite the increasing number of women diplomats since 1972 the atmosphere of the Foreign Office remained paternalistic: the conventional lifestyle expected from individual families was echoed in the make-up of the Service itself. The opinion that working women were inimical to a happy family life was one that was often aired in the DSWA magazine. One article, printed in the Autumn 1987 issue with the ironic title 'The Bad Old Days', was a selectively nostalgic recollection of Foreign Office family past: 'We may have been hard done by, but we were *friends*, and when our children came out for the holidays *they* were friends too. Indeed they thought of themselves as a family, a marvellous ready made "gang"…' (Interesting to note once again that the author felt confident in describing what her children and those of others felt about their holiday experiences.) The article continued in a disparaging tone about changes in the Diplomatic Service and its wives:

> Now that so many wives are working it is becoming more and more difficult to maintain this family spirit. That may not be much of a loss to the young enthusiastic professional woman … by all means let us 'do our own thing' and enjoy the freedom of our own independent way of life and the

freedom of having outside interests of all kinds, but at the same time make sure that we aren't missing out on the fun of a shared embassy social life.[21]

It is essential to note that female diplomats are not mentioned at all in this passage, or in the main article, as though – from the wives' point of view – they derive from a third gender, which is neither 'diplomat' nor 'wife'. The female diplomat appears to have been a difficult figure to assimilate into pre-existing Foreign Office society. McCarthy writes that 'Not until 1987 was a married woman, Veronica Sutherland, appointed to an ambassadorship, closely followed by Juliet Campbell but it was perhaps significant that both had married relatively late in life and were childless'.[22] In contrast, at this time all other major ambassadorial posts in the Foreign Office network were filled by men, although, as the Chief Clerk, Sir John Whitehead, remarked in 1986, '*manpower* as we traditionally call it, the number of *women* in the Service is increasing each year by leaps and bounds'.[23] The way in which the wives related to female diplomats, especially those at higher grades, is difficult to assess. There was also great amount of discussion surrounding the position of male spouses and their position in the Diplomatic Service hierarchy was written about enviously:

> The husbands, it seems, have much greater freedom allowed them to get a job or follow whatever path they may choose. They are not expected or required to cater for cocktail parties, arrange flowers and placement for official dinners ... or even necessarily accompany their wives to official engagements'[24]

Debates about whether to extend membership of the DSWA to the husbands of women diplomats began in 1986 but 'It was not only the reluctance of some to admit the opposite sex, it was also the difficulty of agreeing a name'.[25] Poor response to a ballot on the subject that year meant that no action was taken but by 1990 the DSWA estimated that one in fifteen diplomatic spouses were male: 'The situation has become more acute ... and change is regarded as urgent.'[26] Among alternative names submitted for consideration were the Diplomatic Service Wives and Husbands Association (DSWHA) and the Diplomatic Service Families Association (DSFA) – the second was later adopted and is still in use, but was, in 1990, feared to not represent couples who had no children. Finally committee members settled on the British Diplomatic Spouses Association (BDSA) and the first magazine under the new name was published in 1991: the first time since 1961 that the composition of its members changed, rather than the name of the institution to which they were attached.

A change of attitude

Until late in the twentieth century, young people – especially during their teenage years – were often viewed as an unpredictable force: one to be celebrated but contained. Change happened slowly but as Harrison puts it, in the aftermath of the 1975 Children Act which overhauled the welfare of children in care and outlined the duties and responsibilities of parents and guardians, 'The balance of the law's priorities was shifting away from children's abuse of adults ... towards adults' abuse of children ...'[27] The emergence of children's rights as a concept, forward-looking attitudes towards children and, crucially, a greater awareness of their experiences and anxieties, meant that threatening elements towards children in society were acknowledged. Organisations representing children like Childline, touched on above, provided a means of expression for all young people, aimed to understand the specific problems faced by them and 'no longer assumed that the interests of child and parent coincided'.[28] The first section of this chapter examined the ways in which the institution which governed the lives of Diplomatic Service children was slow to change. However, in the late 1980s sources began to show a greater willingness to discuss difficult issues and gave other – related – indications that change was on the way. This section examines the tensions between a growing awareness of the challenges faced by Diplomatic Service families and the actual changes (in attitude and practice) that began to transform their lives. As recognition of children's agency gathered speed so did an awareness of the sometimes difficult realities of children's lives: this included the uncomfortable fact – demonstrated in the first section by those children who recalled being exposed to areas of dubious practice at boarding school. Government legislation moved to further protect children, but the kind of traditional, private institution where Diplomatic Service children found themselves was not always covered by this. In 1987, for example, corporal punishment was banned in British state schools but it was not until 1999 that it was finally outlawed in independent schools – where the majority of Foreign Office children were educated.

It would be wrong to assume that the DSWA was an entirely conservative body; as we have seen it was at the forefront of a number of family related campaigns through the period discussed in this book but it was unusual for the association to voice direct criticism of the Foreign Office, or the impositions of its way of life on individual families (as we saw in the second chapter, the Foreign Service Association was far more willing to speak frankly). However, a survey of the DSWA magazines for this final period is striking for the frankness with which contributors were beginning to write for the first time. In Spring 1986 the magazine published an anonymous critique of the Foreign Office hierarchy entitled 'Out of Touch' which opined 'I have seen the Service change from good to bad ...' and:

> It is time your magazine published something a little more controversial ... I challenge you to publish my letter and show me that you are at least willing to voice the opinions of wives at the bottom level as well as the top.[29]

In the same year the Autumn issue contained an expressive piece from a diplomats' wife whose son had nearly died after sniffing aerosols at school.[30] The boy was in England and his parents at post in the Caribbean and the article contains poignant details to support the defects in a system that upheld family separation:

> Last Thursday we had a call from our son's housemaster to say that our bright, sensible and sensitive sixteen year old was in hospital having almost killed himself sniffing aerosols. That's hard news to hear when there's no flight available for two days and you're in the West Indies.[31]

The article also drew attention to the private cost of life within institutions that had a high public standing, not unlike life in the Foreign Office:

> If you have sniffed aerosols, you need help. You need counselling and you need support. There are children in the public schools of which we all have a high opinion, who need this counselling and support. Some of them are known to staff, some are not, but few are receiving the care they need. Should the 'reputation' of a school come before our children's welfare?[32]

The anonymous author also directed attention to the kinds of welfare services which the Diplomatic Service, with its emphasis on resilience, would have been reluctant to use or acknowledge in preceding decades:

> Welfare Dept is very clued up on aerosol abuse and can help. I am the first FCO wife to apply to them for such help, although I know for certain that other FCO families have this same problem. (It was another FCO boy who introduced my son to the habit.)[33]

Although the FSWA was originally founded in 1960 on the grounds that diplomats' wives needed somewhere to take their problems, a reluctance to complain which was borne out of a fear of reprisals was something that was expressed in internal literature until the very end of the period under consideration. Diplomats and their wives were anxious that if they 'spoke out' their career would suffer, or they would be the subject of gossip, especially at post. Therefore articles like the one above went a long way to promoting greater freedom of expression. A new-found acknowledgement of the complex problems experienced by diplomatic families and 'exacerbated by life abroad' was linked to this change in attitude. In 1986 a confidential agony column was launched in the DSWA magazine:

> Many Diplomatic Wives suffer their personal crises in silence, held back by the fear of gossip and of affecting their husband's career ... Letters will be answered by a qualified counsellor and will, of course, be treated in complete confidence.[34]

The tone of articles published in the DSWA magazine began to differ conspicuously from self-conscious and provocative critiques that had appeared in the DSWA newsletters in the past, for instance in the exchange of opinions between Margaret Ibbott and Jean Reddaway on the role of the Foreign Office wife. The letter from the wife whose son had sniffed aerosols was also far more frank about the situation in which the whole family had found itself, something totally new to the magazine. Furthermore, the author made the distinction between a regard for institutional reputation and the welfare of an individual child. Articles that appeared in subsequent editions of the magazine demonstrated that many Diplomatic Service families were not only beginning to think in the same way but were no longer shy of saying so. As the 1980s progressed, other articles that gave an honest picture of the complexities of FCO family life, especially for children, began to appear. In Autumn 1988 the magazine printed a short story, 'The Rose Garden', written by a young person who had spent part of their childhood in Afghanistan. The child protagonist of 'The Rose Garden' identified so strongly with Afghanistan as a homeland that she was shocked to discover that her actual 'homeland' England was in a place where she felt completely foreign. Disenchantment with the Diplomatic Service peripatetic lifestyle and its subsequent problems of identity formation are responsible for the child's traumatic return to the UK:

> This child died in an English primary school aged seven-and-a-half. Her bleached hair darkened and the sun seeped from her skin leaving her pale and empty. Her face grew fatter and younger and then one day she looked in the mirror and realised she was someone new, someone English.[35]

In 1989 the experiences of diplomatic families with disabled children were first recognised and written about when an article concerning the short life of Simon, a boy born with Hunter Syndrome, was published, giving a one-off account of overseas life with a special needs child. The article celebrates the often positive experiences of families with small children at congenial posts as the family travelled to Tokyo:

> Despite everything we decided … we should take up our posting, a decision which we were never to regret. Simon thrived on overseas life and the doom and gloom of the diagnosis was pushed to the back of our minds.[36]

The Spring 1990 edition included a short notice about a fund for children with special needs administered by the DSWA Emergency Health and Welfare fund: 'DS parents with disabled children are welcome to write to the Committee with their ideas for putting the money to its best use.'[37]

The autumn of 1990 also saw a story about a Diplomatic Service family's experience of adoption. 'My Darling Daughter' described one family's decision never to tell their daughter that she had been adopted as a baby and how their adopted daughter learned the truth as an adult. Describing those who revealed the secret as 'little foxes… who snapped away at the foundations', the author noted that: 'It was good that my friends and colleagues in the Foreign Service had kept their mouths shut but bad that the "little foxes" had turned out to be members of our own family.' The article has a strange tone, but its significance to this study is not its content so much as its inclusion in the magazine which demonstrates that in order to continue its emphasis on the family, the Diplomatic Service had to accept the many different varieties of family that constituted its personnel and do its best to accommodate them, rather than families accommodating the standards of the Foreign Office, as had happened in the past.

When during the early 1960s civil servants and their wives gave evidence to the Plowden Committee it was in the hope of being able to counteract the threat of real financial hardship; however, by the time a third fare for children's travel was granted in 1971, allowances were viewed by FCO employees with far more of a sense of general entitlement. Always paternalistic, the Foreign Office itself began to be viewed more than ever as a great provider, to be blamed if it was seen to be falling short of providing for its families. Coles felt that the growing system of allowances could lead to '"a tender trap" in which families could find themselves bureaucratically enmeshed'.[38] During this period families were continually pushing for an extension of existing allowances and perceived inequalities in their distribution occupied much of their time. Spring 1988's DSWA magazine documented a number of resentful questions during the Annual General Meeting. The question of wives inability to work was still on the table and the administration was asked why the Service did not provide an allowance to compensate for wives' loss of earnings. One parent with children at school in the French system wondered why did the Service only met the cost of British school fees. The head of a large family thought that baggage allowance should be calculated according to the number of children in the family.[39]

One question posed reflected the actual change – statutory and societal – which was reforming family customs on the national stage. A family asked why a student nurse was not entitled to one fare-paid journey per year. This family was not alone as young people were remaining dependent on their families for far longer, due to a rise in the take-up of higher and further education. Statistics show that the number of first degrees gained in the UK rose from 17,337 in 1950 to 77,163 in 1990.[40] DSWA magazines in the 1980s consistently published letters, articles and advice about the cost of dependent older children on Foreign Office families.

> One of the most serious problems for diplomatic service families is the support of children over 18 who are not in employment. Students between 18 and 21 have one paid journey per year but otherwise they must either stay in England or have their fares paid by their parents. Those who are not students have no fares paid at all.[41]

In 1988 a revised Conditions of Service package 'included provision for all 18–21 year olds who are unmarried and in vocational training or are unemployed to make one CCJ [children's concessional journey] a year. Those in full-time education will in future have two'.[42] The daughters of diplomatic families were in an anomalous position owing to the survival of an outdated rule that supposed they would return to the care of their parents until marriage and which enabled them to travel as dependents.

At the other end of the family lifecycle, the education of preschool children was also under scrutiny. A survey conducted by the DSWA in the late 1980s found that Diplomatic Service parents at post were paying far higher costs for preschool settings than they would be in the UK and the Condition of Service package was adjusted to reflect this in 1989, with the official contribution to preschool fees changed from three days to fifteen hours to allow greater flexibility.[43] A question submitted to the AGM in this year inquired how the plans for a creche at Whitehall were developing and a reasonably full reply was given. As usual, there had been a disappointing response to an initial questionnaire and it was felt that space and cost might present problems with the Whitehall estate although there was a possibility of joining forces with another government department (the ODA was mentioned). A subsidised nursery was opened at the FCDO's King Charles Street site in London in 2001.

Another long-standing Foreign Office prejudice was finally challenged during this period: that of the use of international schools for Diplomatic Service children. Research identified this period as the first which saw families who relied wholly on international schools to educate their children which meant that the whole family was able to remain together at post. In addition, other correspondents who were of school age during this time attended international schools before they left for boarding school later on. Although international schools had existed for a considerable time, especially in large cities like Buenos Aires and Bangkok, and had high standing among the international community, they were not a choice for the children of British expatriates who preferred to send their children 'home' to boarding school. Those parents who did choose an international school setting spoke with enthusiasm about their choice to send their children to international schools at each of their postings. This decision was usually made to consciously obviate family separation, rather than to defy Foreign Office tradition, and did not involve a comparison of the two different styles of education. Parents who opted for international schools felt that sending children away to boarding school undermined their sense and experience of family life. They expressed concern for the children of colleagues that they had seen sent away to school and – for the first time in this survey – questioned the effect of separation on the individual child. However, as will be seen in the following section, this choice did not evade boarding school's traditional elitist connections. The cost of international schools overseas were equivalent to exclusive private schools in Britain and the social networks to which the children were introduced were equally privileged. The choice of international school for their children also had implications for the diplomat's career; the Service had not yet learned to accommodate different choices made by its personnel and some parents had to choose between a job and educational facilities when choosing a posting.

Although the number of contributors to this book who experienced international school were far fewer than those who went to boarding school, overall their recollections were far more positive and reactions to the international school experience were far less ambiguous. Although no two international schools were the same, common characteristics appear to have been a sense of safety and community that verged on the over-protected. During the 1980s Retta Bowen attended the well-known international school in Vienna, established after the Second World War to provide education for the children of military personnel, and remembered the experience as being: 'Really great, really really great. Yeh, I loved it. I loved it.'

According to Harrison, 'the UK's divorce rate, far from stabilising, had, by the mid-1980s, become the highest in Europe', something which 'greatly diversified family shape'.[44] The FCO was no exception to this trend; despite the DSWA magazine feeling able to write 'We may not like to think of divorce as being relevant to ourselves'. By the end of the 1980s the divorce rate among Diplomatic Service couples was one of the highest in the UK and internal literature began to reflect this, publishing sections on marriage and divorce that put frequently asked questions to a female divorce lawyer. The implications of divorce on the diplomatic family were covered in an article in 1990 which adapted typical no-nonsense Foreign Office resilience to this new challenge by ending with the declaration: 'Might I state that I have been there, have even provided Christmas for two young sons on social security, and have survived!'[45] Coles explored the reasons for Foreign Office divorces: 'Many women found that the emotional strain of moving increased over time, not only for family reasons. Spouses might tire of creating a new life with each move: energy and enthusiasm were needed to enter new social circles.'[46] However, of the thirty-three contributors who provided evidence for this book, only two reported family experience of divorce. Retta Bowen commented on the singular set of circumstances that constituted her parents' marriage:

> it was about … maintaining an appearance and my parents had a terrible … marriage, but they were always on show and I think everyone else would have looked at them and regarded them as being 'What a wonderful couple'. Beautiful. 'Aren't they a handsome couple?' and all that sort of nonsense … I think it did force together a family that actually was not together. If they'd had their choice. If everyone had sort of said 'Where do you want to go?' The whole family would have disappeared off in different directions probably.[47]

This sense of enforced interdependence appears to have been common, but at the expense of extended family. This sentiment was echoed by one Diplomatic Service wife who spoke to Coles: 'My husband and I have grown a lot closer over the years. We are the only constant in each other's lives.'[48]

The making of the transnational elite

While families with first-hand experience of international schools – like the ones quoted above – spoke very highly of them, it had never been fashionable for the British Diplomatic Service to show enthusiasm for them or for the ideal of international education. In very much the same way as it remained aloof from post-war suggestions that separation was detrimental to family life, the FCO rejected the international ethos in terms of education. 'Few parents seem to have considered the new international schools', wrote Brendon about the expatriates who stayed on in India in the decade after independence, 'which educated the offspring of ambassadors, businessmen and missionaries from other countries – they were thought to be too American'.[49] The origin of the British belief that international schools were 'too American' is difficult to establish. In all likelihood, it was possibly linked to post-war resentment, as the US supplanted the UK as a world power. There could also have been a connection, within the Foreign Office, to a sense of imperial arrogance that was hard to extinguish. Catherine Webb* recalled her parents' reluctance to send her to the American school when the family was posted to Moscow: 'The only school that was English speaking in Moscow was American and of course it was out of the question for me to go to an American school ...'[50] Evidence that American schools and international schools were viewed as one and the same and that this was a prevalent view among British diplomatic families can be found in a DSWA education report written by a diplomat's wife, Nancy Larmour, who held the DSWA education brief: 'I have almost no experience of American (often called International) schools.'[51] Although this is inaccurate, it shows that the two appear to have been perceived at least as interchangeable. Larmour noted that responses to the questionnaire, by which she gathered information for the report, showed British diplomats and their families 'deploring ... their [the American's] money-weighted influence in English schools' and demonstrating a 'stern disapproval of their methods and discipline, or lack of it'.[52] The Foreign Office attitude towards international schools reflects many of its existing preoccupations: its love of social status, which it felt could only be met in a British public school education, and, perhaps more important, its attachment to tradition and fear of change. Even in the late 1980s British diplomats felt that it was still possible to preserve 'the world apart' that they had carefully cultivated for the large part of the twentieth century.

Writing in 1995, Hayden and Thomson noted that 'the concept of the "international school" is one that has developed rapidly over the past 40 years and is still relatively thinly researched'.[53] They estimated an increase from fifty international schools identified worldwide in 1964 to approximately one thousand at the time they were writing.[54] A 1969 study of international schools written by Robert Leach, a teacher at the International School of Geneva, suggested four categories of international school. The first was the simplest: international schools could be schools that accepted students from a range of national backgrounds; the second was formed of national schools overseas, such as the French school attended by Eleanor King* in South America. Schools that had been founded to represent the interests of more than one nation, like the Anglo-American school attended by Olivia Tate* in Moscow, made up the third category while the fourth group were established to promote international understanding and 'worldmindedness' – to use the remarkable term originated by Sampson and Smith in a 1957 article. According to Sampson and Smith, the 'worldminded' 'favours a world-view of the problems of humanity, whose primary reference group is mankind rather than Americans, British, Chinese etc.'[55]

Hayden and Thompson point out that ideologies such as 'worldmindedness' and others were expressed in guiding principles published by UNESCO and had their basis in the idea of international cooperation as a means of preventing further worldwide conflict in the aftermath of the Second World War. However, this point of view has been challenged by Sylvester, who argues that this is an over-simplification and attempts to illustrate the subtleties in its development between 1893 and 1944.[56]

The gradual adoption of international school as a serious option for British Diplomatic Service children raises questions about the way in which the behaviours of transnational families developed at the end of the twentieth century, especially in terms of the way in which elite identities were revised to suit a new, post-imperial era. A paradigm shift led away from the image of international people as imperial cosmopolitans towards a view of them as 'transnational actors'. Greater willingness among FCO families to use international schools could relate to their development and credibility among the British as the twentieth century progressed, as well as societal changes that, by 1990, characterised separation as synonymous with neglect. Roger Goodman's work on Japan's 'International Youth' and Fiona Moore's on the German School in London suggests that even though the Diplomatic Service parents who chose international schools might have considered them to possess a more democratic and modern culture than boarding schools, international school culture was just as active in seeking to create an elite and that the cost of the education itself was just as expensive as boarding school.[57] However, this was a culture that exhibited different characteristics from those typically found in British public schools and which was more firmly rooted in internationalism and transnationalism. A striking contrast between this elite group of children and boarding school children was the way in which emotional reactions to the features, both positive and negative, of a transient lifestyle and typified in literature on Third Culture Kids and Global Nomads, were more readily encouraged and displayed. These two discourses were unfamiliar to everyone who gave evidence to this book other than the very youngest.

Some of the Diplomatic Service children who attended international school professed a desire to live much of their future life abroad, emphasising the lifestyle enjoyed by Diplomatic Service staff. They also stressed the prestigious nature of the Foreign Office among public sector institutions and the sense of kudos gained from working there and being associated with it. Their background at international schools, coupled with this awareness of status and prestige, illustrated that in terms of identity, these children had distinct ties to what Moore, after Goodman, has termed 'transnational actors', young people 'with an international upbringing who may become the international managers of the future'.[58] Moore's exploration of the pupils and parents of the German school in London considers not only the way in which children at school overseas relate to their host environment and to one another but the ways in which parents work to construct and thus take pride in their children's sophisticated international personas.[59]

'Transnational actor' is perhaps a significant substitute for former identifiers. Eleanor King★ chose to call attention to 'imperial confidence' as an identity which she felt had contributed to diplomats' behaviour and, perhaps more importantly, had become obsolescent:

> you know you understand other people sort of so well and other ways of doing things and … you don't have a culture of your own, you've always been a sort of satellite of another culture and … So you don't really belong anywhere I mean I think in the past people used to say that cosmopolitans were citizens of the whole world. I think there was a certain kind of imperial confidence to that and that these days one is actually much more like sort of a refugee, you actually belong nowhere, as opposed to belonging everywhere.[60]

King★ made a significant point when she compared diplomats and their families to refugees. Literature on transnational movement has largely failed to draw attention to the similarities between privileged migrants who travel overseas on business or government service and less privileged migrants who are forced to travel in pursuit of greater financial security or political refuge, although obviously the comparison is largely superficial. The original edition of Pollock and Van Reken's *Third Culture Kids* (published in 1999) did not include the children of refugees or economic migrants, although similarities between privileged and less privileged child migrants are striking. A revised edition in 2009 modified the definition of 'third culture kids' (TCK) placing it as a sub group within a wider paradigm: 'cross-cultural kids'. The principal difference between these two groups, according to Pollock and Van Reken, relates to the degree and expectation of movement:

> As the cultural mixing of today's world increases, these questions regarding who can or cannot be regarded as an 'official TCK' are important ones to address. Historically we assumed the difference between the TCK experience and that of immigrant children was simple: immigrants moved to a land to stay and many never took even one trip back to the homeland after arriving in the new country. TCKs moved with the expectation of one day returning to their original country.[61]

However, as Kunz has skilfully highlighted, divisions between transnational families are often interpreted in terms of social identity. Scholarly investigation into less privileged migrant groups of non-western origins far outstrips examination of migration experienced by professionals such as business and government employees and aid workers (more often known as 'expatriates') and their families.[62]

An awareness of 'privileged migration' was articulated via the feelings of being 'different' that interviewees often described. The beginning of an 'FCO family turn' towards a different kind of internationalism and elite status during this period suggests an aspiration – in common with other former colonial powers – to revise elite identities in order to suit a new post-colonial era. As Fechter has argued on the subject of 'privileged migration', 'postcolonial dis/connections are thrown into relief most sharply in the case of Euro-Americans moving to developing countries or former colonies ...'[63] In some of the literature on transnationalism there can be a sense that the type of people who typify 'transnational actors' have never existed before. For example, the early pages of Pollock and Van Reken's TCK 'handbook' list the reasons why TCKs have only recently entered the public consciousness. The first quotes Carolyn Smith's 1991 *The Absentee American*: 'Since 1946 ... when it was unusual for Americans to live overseas unless they were missionaries or diplomats it has become commonplace for American military and civilian employees and businesspeople to be stationed abroad, if only for a year.' This passage mentions missionaries and diplomats, two main groups that travel overseas to work, and it is important to note that it only later became 'commonplace' for Americans to form an expatriate culture. By 1946 the concept of living overseas for work and of children moving between cultures had been a long established part of British life among imperial and colonial and diplomatic employees. It could be argued, then, that British diplomatic children – alongside their colonial equivalents – had long been TCKs and Global Nomads, had in fact been their originals, but it is important to recognise that these terms represent a conscious change in terminology and a refreshed perspective. In future studies, the identity of British Diplomatic Service children must be examined alongside a rejection of imperial associations.

Conclusion

This chapter, which dealt with the final time frame comprising this study, from 1985–1990, sought to give an impression of 'changing times' both inside and outside the Diplomatic Service and to effect a comparison of the two. During this era diplomats – in common with the rest of Whitehall – were challenged by drastic cuts to the public sector. Socially, times had changed; the numbers of women diplomats entering the service increased year on year following the lifting of the marriage bar in 1972; divorce was on the rise in the UK and affected the Diplomatic Service particularly; children were 'younger for longer' entering higher education after school. Nonetheless in many areas the Foreign

Office remained slow to change. Some characteristics of its families' lives showed little difference with those described in the first chapter of this book: boarding school was still the norm and children were secondary to the FCO system which was paternalist, a 'big brother'. However, as the second section revealed, societal changes had begun to affect the behaviours of diplomats and their families. Most striking was the willingness to discuss frankly problems like those of marital breakdown and substance abuse that would have been hidden in the past. The DSWA even provided an agony column and qualified counsellor to address the problems faced by wives overseas. The DSWA magazine published articles on a far wider range of topics, including children with disabilities and adoption. The final section examined the way that international schools began to appeal to Diplomatic Service families – for families the most significant break with traditions of the past – and how the international school system, rather than providing a more democratic environment for transnational families, adapted to create a new, post-colonial elite.

Notes

1 Brian Harrison, *Finding a Role: The United Kingdom 1970–1990* (Oxford: Clarendon, 2011), 455.
2 *DSWA Magazine*, Autumn 1988.
3 'Opinion: The Service in Decline', *DSWA Magazine*, Spring 1986, 24.
4 'Chief Clerk's Speech to AGM', *DSWA Magazine*, Autumn 1988, 17–21.
5 Ibid.
6 Peter Clarke, Hope and Glory: Britain 1900–2000 (London, New York: Allen Lane, 2004), 366.
7 Harrison, *Finding a Role*, 242.
8 'Mrs Margaret Morris Child Escort Service', *DSWA Magazine*, Spring 1986, 49.
9 Hereward Harrison, 'Childline – The First Twelve Years', *Archives of Disease in Childhood* 82 (2000): 283–285.
10 See Colin Butler on Childline's proactive approach to technology, in '30 Years of ChildLine (1986–2016)', witness seminar held 1 June 2016, at the BT Tower, London, transcript held at Modern Records Centre, University of Warwick, Coventry, 24.
11 Vyvyen Brendon, *Children of the Raj* (London: Phoenix, 2006), 206–207.
12 'Choosing Boarding Schools', *DSWA Magazine*, Autumn 1990, 101.
13 Interview. Rebekah Lattin-Rawstrone. 21 March 2014.
14 Harrison, *Finding a Role*, 389.
15 Interview. Anne Foster*. 11 February 2015.
16 Ann Coles, 'Making Multiple Migrations: The Life of British Diplomatic Families Overseas', in Anne Coles and Anne-Meike Fechter, *Gender and Family among Transnational Professionals* (London: Routledge, 2008), 143.
17 Zara Steiner, 'The Foreign and Commonwealth Office: Resistance and Adaptation to Changing Times', *Contemporary British History* 18:3 (2004): 22.
18 Ibid., 13–30, 29.
19 Harrison, *Finding a Role*, 456.
20 Katie Walsh, 'Travelling Together? Work, Intimacy, and Home amongst British Expatriate Couples', in Anne Coles and Anne-Meike Fechter, *Gender and Family among Transnational Professionals* (London: Routledge, 2008), 65.
21 Dawn Willson, 'The Bad Old Days', *DSWA Magazine*, Autumn 1987, 67–68.
22 Helen McCarthy, *Women of the World: The Rise of the Female Diplomat* (London: Bloomsbury, 2014), 296.

23 'Chief Clerk's Speech', *DSWA Magazine*, Spring 1986, 16.
24 'Diplomatic Husbands', *DSWA Magazine*, Spring 1986, 89.
25 'AGM Report', *DSWA Magazine*, Autumn 1987, 16.
26 'Editorial', *DSWA Magazine*, Autumn 1990, 7.
27 Harrison, *Finding a Role*, 244.
28 Ibid.
29 Anon., 'Out of Touch', *DSWA Magazine*, Spring 1986, 87–88.
30 'Aerosols: The Hidden Killer', *DSWA Magazine*, Autumn 1986, 100–101.
31 Ibid.
32 Ibid.
33 Ibid.
34 'Editorial', *DSWA Magazine*, Spring 1986.
35 Melanie Birch, 'The Rose Garden', *DSWA Magazine*, Autumn 1988, 66–67.
36 Christine Lavery, 'After Simon', *DSWA Magazine*, Autumn 1989, 73–74.
37 'Children with Special Needs Fund', *DSWA Magazine*, Spring 1990.
38 Coles, 'Making Multiple Migrations', 129.
39 *DSWA Magazine*, Autumn 1987.
40 Figures taken from House of Commons Library Historical Statistics Standard Note:
     SN/SG/4252
         Last updated: 27 November 2012 Author: Paul Bolton
41 *DSWA Magazine*, Autumn 1986.
42 *DSWA Magazine*, Spring 1988, 30.
43 'Conditions of Service Update', *DSWA Magazine*, Autumn 1989, 32.
44 Harrison, *Finding a Role*, 222.
45 Joyce Bentley, 'Divorce and Children', *DSWA Newsletter*, Spring 1990.
46 Coles, *Making Multiple Migrations*, 143.
47 Retta Bowen. Interview by Author. 22 May 2015.
48 Ibid.
49 Brendon, *Children of the Raj*, 266.
50 Interview with Catherine Webb★. 19 August 2015.
51 Larmour Report, 12. Smedley Box B.
52 Larmour Report, 12.
53 Mary Hayden and Jeff Thompson, 'International Schools and International Educa-
     tion a Relationship Reviewed', *Oxford Review of Education* 21:3 (1995): 327–345.
54 Hayden and Thompson, 'International Schools', 332.
55 Donald L. Sampson and Howard P. Smith, 'A Scale to Measure World-Minded
     Attitudes', *The Journal of Social Psychology* 45:1 (1957); 99–106.
56 Robert Sylvester, 'Mapping International Education: A Historical Survey 1893–
     1944', *Journal of Research in International Education* 1:1 (2016): 90–125.
57 Roger Goodman, Japan's International Youth: The Emergence of a New Class of
     Schoolchildren (Oxford: Clarendon: 1990); Fiona Moore, 'The German School in
     London, UK: Fostering the Next Generation of National Cosmopolitans', in Coles
     and Fechter, *Gender and Family*.
58 Moore, 'The German School in London', 84; Goodman, *Japan's International Youth*.
59 Moore, 'The German School in London', 85–103.
60 Interview. Eleanor King★. 11 July 2014.
61 David Pollock and Ruth E. Van Reken, *Third Culture Kids: Growing Up Among
     Worlds* (London: Nicholas Brealey Publishing, 2009), 30.
62 Sarah Kunz, 'Privileged Mobilities: Locating the Expatriate in Migration Scholar-
     ship', *Geography Compass* 10:3 (2016): 89–101.
63 Anne-Meike Fechter and Katie Walsh, 'Examining "Expatriate" Continuities:
     Postcolonial Approaches to Mobile Professionals', *Journal of Ethnic and Migration
     Studies* 36:8 (2010): 1197–1210.

Bibliography

Beryl Smedley Archive. Archive Box B: Press cuttings and FSWA/DSWA material relating to diplomatic wives.

Brendon, V. *Children of the Raj* (London: Phoenix, 2006).

Clarke, P. *Hope and Glory: Britain 1900–2000* (London, New York: Allen Lane, 2004).

Coles, A and Fechter, A. eds. *Gender and Family Among Transnational Professionals* (London: Routledge, 2008).

DSWA Newsletter Autumn 1987; Autumn 1988; Autumn 1989; Autumn 1990; Spring 1986; Spring 1988; Autumn 1990.

Fechter, A.M. and Walsh, K. 'Examining "Expatriate" Continuities: Postcolonial Approaches to Mobile Professionals', *Journal of Ethnic and Migration Studies* 36:8 (2010): 1197–1210.

Goodman, R. *Japan's International Youth: The Emergence of a New Class of Schoolchildren* (Oxford: Clarendon: 1990).

Harrison, B. *Finding a Role: The United Kingdom 1970–1990* (Oxford: Clarendon, 2011).

Harrison, H. 'Childline – The First Twelve Years', *Archives of Disease in Childhood* 82 (2000): 283–285.

Hayden, M. and Thompson, J. 'International Schools and International Education a Relationship Reviewed', *Oxford Review of Education* 21:3 (1995): 327–345.

Kunz, S. 'Privileged Mobilities: Locating the Expatriate in Migration Scholarship', *Geography Compass* 10 (2016): 89–101.

McCarthy, H. *Women of the World: The Rise of the Female Diplomat* (London: Bloomsbury, 2014).

Moore, F. 'The German School in London, UK: Fostering the Next Generation of National Cosmopolitans', in *Gender and Family Among Transnational Professionals*, ed. A. Coles and A. Fechter (London: Routledge, 2008).

Pollock, D. and Van Reken, R. *Third Culture Kids Growing Up Among Worlds* (London: Nicholas Brealey Publishing, 2009).

Sampson, D. and Smith, P. 'A Scale to Measure World-Minded Attitudes', *The Journal of Social Psychology* 45:1 (1957): 99–106.

Steiner, Z. 'The Foreign and Commonwealth Office: Resistance and Adaptation to Changing Times', *Contemporary British History* 18:3 (2004): 13–30.

Sylvester, R. 'Mapping International Education: A Historical Survey 1893–1944', *Journal of Research in International Education* 1:1 (2016): 90–125.

'30 Years of ChildLine (1986–2016)', Witness seminar held 1 June2016.

Walsh, K. 'Travelling Together? Work, Intimacy, and Home amongst British Expatriate Couples in Dubai', in *Gender and Family Among Transnational Professionals*, ed. A. Coles and A. Fechter (London: Routledge, 2008).

Conclusion

Chapter One of this book about Diplomatic Service families and the mobile lives of their children begins in 1939 with an account of the attempts by one diplomatic grandee to exclude members of the Consular Service, the 'Cinderella Service' from diplomatic circles, by opposing a suggested amalgamation of the two Services. The letter from Sir Hughe Knatchbull-Hugessen to the Permanent Under Secretary (PUS) Sir Alexander Cadogan introduced many of the themes that become familiar over the course of the narrative. One is the way that the Diplomatic Service clung to ideas of elite social status. Another is the characteristic ambiguity with which these ideals were introduced. The enduring nature of these tenets can be traced right up to the final chapters, which covered the status-conscious 1980s and demonstrated that diplomats retained the sense of belonging to a prestigious social group well into that period. Even when the outmoded, establishment practices which they held dear finally began to become untenable, this sense of superiority was transmogrified via an identification with an emerging transnational elite. Ambiguity was another long lasting characteristic of Diplomatic Service practice, the unofficial service demanded from Foreign Office wives, again an expectation well into the 1980s, had no formal basis and yet to challenge it was to mark yourself and your family out as awkward and to possibly injure your husband's career.

The dominant theme of this historical account of the Foreign Office family experience in the late twentieth century has been change, or, rather, the lack of it. Chapter Two, with its accounts in the first section of the evidence given to the Committee on Representational Services Overseas, 'the Plowden Report' and in the third section of internal lobbying for a third 'concessionary' journey to enable children to travel to and from overseas posts and the granting of many of these demands might point to significant internal change. However, as Margaret Ibbott pointed out in her letter to the Diplomatic Services Wives Association (DSWA) magazine about the obligations of Foreign Office wives, it was all about 'attitude'. This intangible notion, hard to grasp and describe, is responsible for much of the lack of change that moved in tandem with the kinds of small victories described above. The description of family separation contained in the third section of Chapter One is hardly changed from its revisitation in the first section of Chapter Four. There were good reasons why

DOI: 10.4324/9780429273568-5

boarding school was preferable for British Diplomatic Service families; they were outlined by Peter Boon in the early section and Anne Foster⋆ in the later one. But the reason why boarding school and family separation endured into the early 1990s was down to tradition: things had always been done that way. The Foreign Office, with its strong identity as a 'total institution', appeared to be more comfortable with similar establishments, which can be seen in its reoccurring emphasis on the use of the British public school system. Additionally, the attractive public image of the Diplomatic Service, as luxurious, elite and exciting, began to feed on itself in the way that Harrison describes the television programme *Yes, Minister* influencing the behaviours of civil servants in other government departments. Conservatism in 'attitude' – the unlikelihood of rebellion among Diplomatic Service children – was linked to this unwillingness to change. Individuals attracted to service life were conservative by nature, had links to other service or overseas backgrounds (the military or missionary life) or were 'following in their father's footsteps' to recall Steiner's phrase. It is reasonable to suppose that these individuals found spouses with a similar outlook and that the family culture, which influenced their children, was again similar.

The third section of Chapter Three, which dealt with the increased level of threat suffered by Diplomatic Service families overseas, demonstrates that this resistance to change could put families in danger. Retta Bowen's description of her family's experiences in Lusaka during wartime confirmed that 'home' for diplomats overseas was not a haven and raised the question why her father was accompanied into a very dangerous situation, with the very real threat of sexual violence, by his wife and two young daughters. The death of Christopher Ewart-Biggs in the Republic of Ireland in July 1976 was an extreme and rare occurrence but the presence of his family at his place of work which was also his home rendered the incident all the more poignant and traumatic. The burden placed on Jennifer Hickman, the wife of Ewart-Biggs' 'number two' at the Dublin Embassy, who had to take charge of the Ewart-Biggs household immediately after the bereavement and who was prevailed on to tell the older children of their father's death, seemed immeasurable. It indicated that no formal system was in place in the event of incidents such as these. Hickman, her daughter and the older Ewart-Biggs children were expected to draw on the resilience and sang-froid that the Diplomatic Service prided in its members, although, again, nowhere was the acquirement of this attribute formally given or detailed. This lack of formal preparation was what angered Julia Miles, wife of the ambassador in Libya, who was evacuated from Tripoli with her young family and other wives and their families in April 1984. She arrived in London to discover that Foreign Office representatives had not even known that they were able to contact her by telephone in Libya which was why no one had contacted the family to ask after their wellbeing. Miles and the other wives from Tripoli were not offered any help and relied on family and friends for accommodation and care when they arrived in England.

Resistance to change in twentieth century Foreign Office society also meant that the entrenched gendered aspects of Diplomatic Service life were maintained for far longer than they were outside the organisation. Throughout the 1970s Foreign Office communities of male diplomats were assisted by their wives who undertook unpaid and unrecognised work related to the Service. A new generation of wives began to question this requirement and support and acknowledgement of their position came through two academic works, Callan's *Premiss of Dedication*, in 1976 and *The Incorporated Wife* in 1984, which Callan edited with Ardener, another social anthropologist. A question crucial to this book was whether Diplomatic Service children felt that they were 'incorporated' into the Service in the way that their mothers were but although their lives were largely dictated by their father's membership of the Foreign Office they did not feel the same allegiance to it.

The Diplomatic Service has always been a paternalistic institution. Accounts of the evacuations from Libya indicated how separate were the spheres occupied by the male diplomats and their wives. When the marriage bar was lifted for women diplomats in 1972, the delicate balance within the Foreign Office of male diplomats 'working' alongside their wives was challenged. But the Diplomatic Service, with its traditional values, has always favoured men and comments from diplomatic wives suggested that there were far fewer expectations of diplomatic husbands who appeared to be able to seek the type of work they wanted outside the embassy and who were not required to produce dinners and organise cocktail parties in the same way. The changing of the name Diplomatic Service Wives Association to the British Diplomatic Spouses Association in 1990 indicated that real change was imminent. It is also interesting to speculate whether the majority of male diplomats in the earlier years of this study led to an attachment to traditional establishment values and whether this began to shift when larger numbers of women diplomats became prominent in the Service. It is important to note, however, that notions of the Service as an elite body did not disappear but adapted to a more suitable international status in a post-colonial setting.

A further integral theme of this study is that British diplomats and their families were largely at odds with the country that they were required to represent overseas. This was partly because parents spent so much time outside it, making, in Coles' words, 'multiple migrations' to and from the UK, and partly because children, most often in UK boarding schools, shared an unusual experience with a diminishing group of peers. In the early stages of this book's time frame, the experiences of Diplomatic Service children began to diverge from those of their UK contemporaries who reaped the benefits of 'child-centred' post-war policies, although they shared many characteristics with upper class children and those of colonial administrators. The growing gulf in experience between Diplomatic Service children and their UK-based counterparts was highlighted in the first section of the final chapter by the appearance of an advert for informal childcare services with an apparently unregistered provider in the same year as the children's charity *Childline* was founded. By the very end of this study numbers of children at boarding school nationally had fallen: tying Diplomatic Service children unequivocally to their experiences and identity through their attachment to the Service.

Bibliography

Callan, H. 'The Premiss of Dedication: Notes towards an Ethnography of Diplomats' Wives', in *Perceiving Women*, ed. S. Ardener (London: Malaby Press, 1975).

Callan, H. and Ardener, S. eds. *The Incorporated Wife* (London: Croom Helm, 1984).

Coles, A. and Fechter, A. eds. *Gender and Family Among Transnational Professionals* (London: Routledge, 2008).

Diplomatic Service Wives Association (DSWA) Magazine, Spring 1976.

Evidence submitted to the Committee on Representational Services Overseas (The Plowden Committee). Bound volumes: Committee on Representational Services Overseas: Evidence RSO (62) 1–30.

Evidence submitted to the Committee on Representational Services Overseas (The Plowden Committee). Bound volumes: Committee on Representational Services Overseas: Evidence RSO (62) 31–59.

Evidence submitted to the Committee on Representational Services Overseas (The Plowden Committee). Bound volumes: Committee on Representational Services Overseas: Evidence RSO (63) 1–12.

Evidence submitted to the Committee on Representational Services Overseas (The Plowden Committee). Bound volumes: Committee on Representational Services Overseas: Evidence RSO (63) 13–59.

Harrison, B. *Finding a Role: The United Kingdom 1970–1990* (Oxford: Clarendon, 2011).

Platt, D. *The Cinderella Service: British Consuls since 1815* (London: Longman, 1971).

Steiner, Z. 'The Foreign and Commonwealth Office: Resistance and Adaptation to Changing Times', Contemporary British History 18:3 (2004): 13–30.

Appendix 1: Interviewees

Olivia Tate★

Born	1946, West Midlands.
Parents	Father, HM Diplomatic Service; Mother, housewife.
Siblings	One brother, one sister.
Occupation	Interior Designer.
Interview location	Tate★ home, London.

Rebekah Lattin-Rawstrone

Born	1976, Kenya.
Parents	Father, HM Diplomatic Service; Mother, housewife, former Civil Servant.
Siblings	One sister, one brother.
Occupation	Writer.
Interview location	Author's home, London.

Peter Boon

Born	1942, India.
Parents	Father, HM Diplomatic Service; Mother, Housewife.
Siblings	One brother.
Occupation	HM Diplomatic Service (retired).
Interview location	Boon home, Oxfordshire.

Eleanor King★

Born	1954, Istanbul.
Parents	Father, HM Diplomatic Service; Mother, Housewife.
Siblings	Two sisters.

Occupation   Writer, Editor, Translator.
Interview location   King* home, London.

Kate Howells

Born   1962, Sussex.
Parents   Father, HM Diplomatic Service; Mother, Housewife.
Siblings   One brother.
Occupation   Journalist, BBC World Service.
Interview location   Howells home, London.

Julia Miles

Born   1945, Oxford.
Parents   Father, Academic; Mother, profession unknown.
Siblings   Unknown.
Occupation   Counsellor and Psychotherapist.
Interview location   Miles home, Oxford.

Oliver Miles

Born   1936, London.
Parents   Father, formerly Ceylon Civil Service; Mother, unknown.
Siblings   Unknown.
Occupation   HM Diplomatic Service (retired).
Interview location   Miles Home, Oxford.

Paul Tylor

Born   1964, Hertfordshire.
Parents   Father, FCO Security Officer Cadre; Mother, housewife.
Siblings   Two brothers.
Occupation   HM Diplomatic Service.
Interview location   Foreign and Commonwealth Office, London.

Vicky Tarry

Born   1986, Paris.
Parents   Father, HM Diplomatic Service; Mother, Housewife.
Siblings   none.
Occupation   Home Civil Service, FCO.

Born	1986, Paris.
Interview location	Tarry home, Hertfordshire.

Anne Foster★

Born	1945, Wiltshire.
Parents	Unknown.
Siblings	Two brothers.
Occupation	HM Diplomatic Service (retired).
Interview location	Foster★ home, London.

Kate Morris★

Born	1964, Hampshire.
Parents	Father, HM Diplomatic Service; Mother, Housewife.
Siblings	Two sisters.
Occupation	Arts administration.
Interview location	Morris★ Office, London.

Andrew Graham

Born	1956, London.
Parents	Father, HM Diplomatic Service; Mother, Housewife.
Siblings	One brother, one sister.
Occupation	Former soldier.
Interview location	Royal Geographic Society, London.

Retta Bowen

Born	1977, Oxford.
Parents	Father, HM Diplomatic Service; Mother, Housewife, Actor.
Siblings	One sister.
Occupation	Psychotherapist.
Interview location	Bowen home, London.

Catherine Webb★

Born	1958, Jordan.
Parents	Father, HM Diplomatic Service; Mother, Housewife, Teacher.
Siblings	One brother, one sister.
Occupation	Teacher.

Born   1958, Jordan.
Interview location   Webb* home, Oxford.

Appendix 2: Little friends of all the world? The experiences of children in the UK Diplomatic Service 1945–1990

Questionnaire

1   To begin, a very general question. What, in your opinion, were the best and worst things about British diplomatic family life? What 'overall' impressions did it leave you with?

2   Could you briefly describe your/your spouse's career, their roles and the postings you/they went on?

3   What were the ways in which your 'Britishness' as a family was emphasised to you? How 'British' did you feel. How 'British' do you feel now?

4   How significant was the Foreign Office in your family life, did you feel that it influenced every family member?

5   What are your feelings about the word 'home'? Where do you consider 'home' to be? How did you create/maintain an atmosphere of home in each overseas post? Were there familiar objects that travelled with you and helped to 'orient' the family?

6   Could you describe your child's education? What were your feelings about boarding school and the traditions that uphold it? What did your child/ children do during the school holidays?

7   Did your child/children ever rebel? Were they involved in any youth subcultures? Did the 1960s counterculture influence you?

8   How aware were you of rank within the service?

9   How far do you feel you experienced life in each host country? Did you learn languages? Did you form lasting friendships with people native to that country? How far did the culture influence you?

10   What are your feelings about the FCO now? Are you still in touch with FCO life?

11   **For diplomats' spouses:** Did you feel that you had a role? Did you feel, as a diplomat's spouse, that you had a special 'duty' which spouses of other people with international careers didn't have, to represent your country?

12   Were you or your family ever in danger?

13   Have you ever felt that the FCO had a very specific influence on your family life which may have led to patterns of behaviour i.e. alcoholism, marital disputes etc.?

14   If you have any stories or observations you'd particularly like to tell please add them! Also – if you have any documents or photographs you could scan and send to me I would be very pleased to have them.

Index

Printed in the United States
by Baker & Taylor Publisher Services